JOHN DEE ON ASTRONOMY

John Dee on Astronomy

Propaedeumata Aphoristica (1558 and 1568),
Latin and English

Edited and translated, with general notes, by

WAYNE SHUMAKER

With an introductory essay on Dee's mathematics and physics
and his place in the scientific revolution by

J. L. HEILBRON

University of California Press, Berkeley, Los Angeles, London

University of California Press
Berkeley and Los Angeles, California

University of California Press, Ltd.
London, England

ISBN 0–520–03376–0
Library of Congress Catalog Card Number: 76–50254
Printed in the United States of America
Designed by Jim Mennick

1 2 3 4 5 6 7 8 9

❧ Contents

❧ Preface

Our purpose is to make John Dee's *Propaedeumata Aphoristica* (1558–1568) accessible to modern readers and to relate it to Dee's career and to intellectual history, specifically, to the history of science.

Apart from the brief *Parallaticae commentationis praxeosque nucleus quidam* (1573), all of Dee's published works now become available in English. The other major Latin work, the *Monas hieroglyphica* (1564), has been twice translated into English in recent years, by J. W. Hamilton-Jones in 1947 and by C. H. Josten in 1964. It and the *Propaedeumata*, together with the "Mathematicall Preface" and addenda to Sir Henry Billingsley's translation of Euclid's *Elements* (1570), make up Dee's published contributions to natural philosophy. The *Monas* was not understood when published and remains largely opaque. In general, discussions of Dee as mathematician and scientist have hitherto drawn very heavily upon the "Mathematicall Preface" but hardly at all upon the Euclidean addenda or the *Propaedeumata*, which have equal or greater importance. We hope that this imbalance can now be rectified.

We have tried, through the "Introduction" and "General Notes," to make the *Propaedeumata* comprehensible in and for itself. This is not easy, for much is not immediately intelligible to readers whose preassumptions and education differ enormously from the author's. Besides changing the language (though the original Latin is retained on facing pages) and attempting to penetrate the "veiled utterance" traditional among occultist writers, we have brought to bear upon the text other pertinent information about Dee's life, his other writings, and the intellectual habits of the mid-sixteenth century. Although some difficulties remain unresolved (for instance, the baffling initials of Aphorism XVIII), the major obstacles have been surmounted. Our interest in relating Dee to his period has had, however, a second aspect.

Historians have long disputed about the relation between the Renaissance and the scientific revolution of the seventeenth century. The prevailing view in the late nineteenth century was that the Renaissance dispelled the fog of scholasticism and prepared unprejudiced minds to receive the clear light of science. A generation later, when Pierre Duhem and those he inspired came to read the scholastics, they found much that sounded like Galileo; and it began to appear that the Renaissance, rather than nourishing the scientific revolution, delayed its appearance for two hundred years. This interpretation was challenged in the 1930s by both social and intellectual historians. According to the former, the Renaissance created protoscientists in the form of artist-engineers, improving artisans, surveyors, navigators, and other applied mathematicians, a miscellaneous class of semilearned, practical men, disdainful of the disputations of the schools and eager for a science capable of improving man's estate. According to the latter, the Renaissance spawned Pythagoreans, Neoplatonists, numerologists, and cabalists—number-magicians who inspired the scientific revolutionaries to build a quantitative, anti-Aristotelian physics. The latest proposal unites these schools of thought and looks to Renaissance applied mathematicians inspired by Platonizing or occultist philosophies as the true and necessary precursors of the scientific revolution.

John Dee is the exemplar of both sorts of precursors. An

occultist who talked to angels, a mathematician consulted by navigators and geographers, a mystical alchemical adept and a plodding teacher of arithmetic, Dee has become a test case for the new interpretation of the Renaissance roots of the scientific movement. Dee's admirers emphasize his mathematics and the fruitful tendencies of his hermeticism; his detractors point to his unintelligible hieroglyphic monad, his chats with angels, and an insufferable vanity that duped both himself and his latter-day admirers into inflating his importance. In fact the major texts—the *Propaedeumata* (first published in 1558), the *Monas* (1564), the "Mathematicall Preface" (1570), and the records of angelic conversations and alchemical experiments (beginning in 1581)—show a continuous progress toward the occult and the irrational. The first of the series, the *Propaedeumata*, is, in the main, a fully intelligible series of recipes for applying arithmetic and geometry to a standard scholastic physics and astronomy. The implied conclusion is that Dee's mathematics did not grow from or together with his occultism but, rather, preceded it, and that in so far as he devoted himself to occult studies he moved off the high road of the scientific revolution.

It remains to indicate the division of work between the collaborators. Wayne Shumaker established the Latin text and made the translation, to which he has added "General Notes." John Heilbron wrote the "Introduction," references to which are included among the "General Notes." Each has profited from the comments of the other. The "Preface" has been written jointly.

Berkeley, California
August 5, 1976

INTRODUCTORY ESSAY

by J. L. Heilbron

≋ Abbreviations Used
in Footnotes

AT	John Dee. *Autobiographical Tracts*. Ed. James Crossley. Manchester, 1851.
DNB	*Dictionary of National Biography*.
DSB	*Dictionary of Scientific Biography*.
"Preface"	John Dee. "Mathematicall Preface." In Euclid, *Elements of Geometrie*. Ed. Henry Billingsley. London, 1570.
	Reprinted as *The Mathematical Preface to the Elements of Geometry*. Ed. Allen G. Debus. New York, 1975.
Thorndike	Lynn Thorndike. *History of Magic and Experimental Science*. 8 vols. New York, 1923–1958.

✎ 1. Dee's Role in the Scientific Revolution

I. DEE'S CAREER

John Dee (1527–1608) was a geometer, physicist, astrologer, antiquarian, hermetist, and conjurer, a mixture of mathematician and magician, of scholar and enthusiast, of schemer and dupe. He began to acquire his stock in trade, great erudition and a reputation for it, at St. John's College, Cambridge, where he was an undergraduate from 1542 to 1545. "I was [then] so vehemently bent to studie," he later wrote, "that for those yeares I did inviolably keepe this order: only to sleepe four houres every night; to allow to meate and drink (and some refreshing after) two houres every day; and of the other eighteen houres all (except the tyme of going to and being at divine service) was spent in my studies and learning."[1] Dee drew up this account of his undergraduate labors in 1592, as part of a *Compendious Rehearsal* of his achievements and sacrifices in the cause of learning; he hoped thereby to refute the charge of conjuring often brought against him, and to persuade Queen Elizabeth that he merited a place or pension from a grateful nation. One would have to be more gullible than Dee himself to credit his *Rehearsal* literally; but one

1. Dee, *AT*, p. 5. Most of what is known about Dee's life is summarized in *DNB*, V: 721–729.

would correctly infer from it that, from the beginning, he based his career on knowing, or appearing to know, more than anyone else.[2]

In Dee's Cambridge days the academic fad was Greek. He studied it vigorously, probably under the guidance of John Cheke, Fellow of St. John's and Regius Professor of Greek, then engaged in the exciting and dangerous attempt to impose a new pronunciation upon the members of the university. Although neither Cheke nor Dee was a great Grecian, each drew much from his study of the language. In 1544 the professor became tutor in ancient languages to the future Edward VI, and two years later the student received a foundation fellowship and appointment as under-reader in Greek at Trinity College, Cambridge, then newly established by Henry VIII.[3]

The Court already knew something about Dee. His father, Rowland Dee, a mercer by trade, was a "gentleman sewer," or maître d'hôtel, to Henry, from whom he had received several tangible marks of favor. In 1544, for example, Rowland Dee began to enjoy a curious monopoly in the freight-forwarding business: the right to supervise, for a fee, the packing of all merchandise shipped abroad from London. This plum was sweetened on at least two occasions by the award of income from confiscated Church lands.[4] It came naturally to Rowland Dee's son, who later tried to trace his lineage to the ancient kings of Wales, to center his expectations on princes. John Dee was a courtier by nature and nurture, as the eugenicists used to say. Eventually he married into the group of high-placed hangers-on. His third wife was the daughter of a lady-in-waiting. He had begun with the widow

2. For other pertinent examples of upward mobility conferred by reputation for learning, see J. Burckhardt, *Civilization* (1958), I: 27–28, 229–235; Nauert, *Agrippa* (1965), p. 33.

3. *AT*, p. 5; J. Strype, *Life of Cheke*, 2nd ed. (1821), pp. 12–13; A. Tilley, *Engl. Hist. Rev.* 53 (1938): 445–454.

4. I. R. F. Calder, *John Dee* (1952), II: 90–91.

of a London grocer, perhaps a purveyor to the royal house-
hold, precisely the sort of person with whom Rowland Dee
would have associated.[5]

Dee's ambitions extended far beyond the reach of a don.
Even in respect of learning, neither English university could
contain him. However excellent in Greek and Divinity, they
had little to offer in mathematics, in which Dee early had an
interest, and, if we credit the "Preface," a considerable skill:
"in my youth," he wrote there, "I inuented a way, How in
any Horizontall, Murall or Aequinoctiall Diall, etc., At all
howers (the Sunne shining) the Signe and Degree ascendent,
may be knowen."[6] (What this signifies will be made clear
later.) Dee doubted that he could go much further unaided;
and since, in his judgment, no one at Oxbridge, indeed no one
in England, was prepared to "set furth his fote or shew his
hand" in the higher mathematical sciences,[7] he had perforce
to visit the Continent.

His quest for foreign knowledge began in the Low
Countries in 1547. He found what he wanted, and the following
year, after bringing back to England some excellent astronomical instruments, he returned for protracted study at the
University of Louvain. The institution was then at its prime.
It had financial support from the Pope and the Emperor
Charles V; its student body, drawn from many parts of
Charles's vast dominions, was reckoned the largest in Europe
outside Paris; and its faculty, especially of civil law and
mathematics, had acquired an international reputation.[8] Dee
studied both subjects, chiefly mathematics, in which he had
the guidance of Gemma Frisius, professor of medicine, and of

5. C. F. Smith, *John Dee* (1909), pp. 13, 46; E. G. R. Taylor, *Tudor Geography* (1930), pp. 75–76, 107.

6. "Preface," sig. d.ij^r. Dee's invention apparently showed the rising point of the ecliptic; see chap. II: 3 below. For the few mathematical students at Cambridge circa 1550, see Calder, *Dee*, I: 198; II: 99.

7. Dee to Cecil, 1563, cited by Calder, *Dee*, I: 198.

8. L. van der Essen, *Rev. gén. belge*, no. 43 (May, 1949): 38–39, 57; Louvain University, *L'Université* (1900), pp. 10–16.

Gemma's student, Gerard Mercator, already a well-known cartographer. These gentlemen were more than fountains of technical information. They showed where mastery of Renaissance mathematics might lead. Both were welcome at the imperial court at Brussels, where they advised on geographical matters and did a good business in globes and other mathematical furniture.[9]

"Mathematics" in Dee's time included many subjects that have since become branches of physics or engineering, or entirely separate disciplines: optics, architecture, surveying, fortification, cartography, astronomy, navigation. The association of mathematics with practical application gave philosophers who did not understand it a pretext to despise it, and well into the seventeenth century mathematics was often depreciated for its odor of practicality.[10] Consequently, Renaissance mathematicians sometimes sought to legitimize their study by reference to Platonizing philosophies that gave numbers special powers, raised geometrical proof to a level just beneath God's own way of reasoning, and made numerical relations the blueprint for the construction and maintenance of the universe. In some cases, appeal to Plato or Pythagoras might be little more than an academic tactic or a literary flourish.[11] In others, Pythagoreanism may have provided the motivation and, more rarely, the guide, to mathematical

9. P. Gilbert and B. Lefebvre, *Rev. quest. sci.* 12 (1927): 20–21.

10. P. Allen, *J. Hist. Ideas* 10 (1949): 219–253; C. J. Scriba in Roy. Soc. of London, *Notes and Records* 25 (1970): 17–46; J. Webster, *Academiarum examen* (1654), pp. 41–42; Calder, *Dee*, I: 288, citing Lord Herbert of Cherbury, *A Dialogue between a Tutor and his Student* (London, 1768), p. 2: the ends of mathematics are "ignoble . . . as tending only to the measuring of heights, depths and distances," which, however useful, "can be in no way esteemed, as objects adequated or proportioned to the dignity of our souls."

11. For example, Ramus writes in the preface to his Euclid (1544): " . . . minime mirandum est Pythagoram, Platonemque huius admiratione disciplinae [i.e., mathematics] captos, maius in ea diviniusque quiddam deprehendisse, quam ut humanis sensibus tribuendum arbitrarentur." Ramus, *Praefationes* (1599), p. 120.

work.[12] The Louvain group, despite a strong interest in astrology, did not draw its inspiration from Pythagoras. Initially, Dee identified himself with them and with other leading applied mathematicians, particularly the hard-headed Portuguese cartographer Pedro Nuñez (1502–1578), professor of mathematics at the University of Coimbra, chief technical advisor to the King of Portugal, scourge of circle-squarers and of fuzzy-minded astrologers.[13] Gradually, however, Dee's sympathy for Platonizing philosophies intensified, and he ended closer to the Neoplatonists Ficino and Pico della Mirandola than to Nuñez and Gemma Frisius.

From Louvain, Dee made his way to Brussels and then to Paris, where he sought out the mathematicians: Petrus Montaureus (Pierre Mondoré), the King's librarian, who lent him a copy of Diophantus; Oronce Finé, professor of mathematics, astrologer, cartographer, and circle-squarer, to whose "impudent, false and ignorant" claims Nuñez had devoted an entire book;[14] Antoine Mizauld (Antonius Mizaldus), who styled himself "physician and mathematician," and taught medical astrology; the famous Pierre de la Ramée (Petrus Ramus); and a dozen more. Dee began to demonstrate his prowess before his new acquaintances in September 1550, in well-attended lectures on the first two books of Euclid's *Elements*.[15] So great a curiosity as an Englishman who understood geometry deserved attention. Dee was advertised and encouraged by

12. For example, Joannes Caeserius, epistle dedicatory to a collection of mathematical tracts (1507): "[numbers] non esse res leves aut omnino vulgares (quamquam et vulgus quoque numerare novit), sed plenos divinorum mysteriis et divina quadam ratione. . . ." E. Rice, ed., *Prefatory Epistles* (1972), p. 171.

13. Nuñez, *De crepusculis* (1542), letter dedicatory; *DSB*, X: 160–162; F. Gomes Teixeira, *História* (1934), pp. 100–190.

14. Dee, "Preface," sig. *ij^r, and *AT*, p. 8; R. P. Ross, *Studies on Finé* (1971), pp. 262–263; Nuñez, *De erratis Promtii* [!] *Finaei* (Coimbra, 1546); Thorndike V: 299–301.

15. Dee gave the title of the lectures, now lost, as "Prolegomena et dictata Parisiensia in Euclidis Elementorum Geometricorum librum primum et secundum," in *AT*, p. 25.

several fellow countrymen, notably Sir William Pickering (1516–1575), a diplomat with a strong interest in applied mathematics, who had engaged him as a tutor in Louvain and remained a friend.[16] Young Dee's willingness to set forth Euclid in the stronghold of men like Ramus, Finé, and Montaureus, each of whom had either published or had in hand an edition of the *Elements*, argues both competence and courage; his work with Gemma had added information to the confidence he never lacked.

Dee made much of this episode in the *Compendious Rehearsal*:[17]

> At the request of some English gentlemen, made unto me to doe somewhat there [Paris] for the honour of my country, I did undertake to read freely and publiquely Euclide's *Elements Geometricall, Mathematicè, Physicè, et Pythagoricè*; a thing never done publiquely in any University of Christendome. My auditory in Rhemes College [the insignificant Collège de Reims] was so great, and the most part elder than my selfe, that the mathematicall schooles could not hold them; for many were faine, without the schooles at the windowes, to be auditors and spectators, as they best could help themselves thereto. I did also dictate upon every proposition, beside the first exposition.

Now Dee had lectured on only the first two of Euclid's thirteen books, not upon "every proposition," and his treatment was by no means unprecedented. To speak only of Paris, Finé had been lecturing on Euclid publicly and mathematically for many years, and the famous Jacques Lefèvre d'Étaples had taught how to read him "pythagoricè."[18] It was already a commonplace to refer to Pythagoras, Plato, and

16. Ashmolean MS 423, ff. 294–295 (Bodleian); J. W. Burgon, *Life of Gresham* (1839), p. 459; *DNB*, XV: 1130–1131; Dee to Commandino [1563?], in Euclid, *Elements*, ed. Leeke and Serle (1661), p. 608: "[that] singular patron of all good Arts, and specially of the Mathematicall, Sir William Pickering Knight, my exceeding good friend."

17. *AT*, p. 7.

18. Ross, *Studies*, Chap. 6.

the sublime when recommending the study of the *Elements*. Everything necessary had long lain ready to hand in prolix Proclus's *Commentaries* on Euclid's first book: by study of mathematics the soul can climb from Plato's cave into full light, awaken from its slothful sleep, gain clear knowledge, become perfect, "attain the happy life by the discovery of pure thought."[19] In his edition of Euclid published in 1516, Lefèvre endorses mathematics as the Greek way to the contemplation of divinity; "it has no trace of the filthy, of anything carnal," he says, "and it was the most ancient way of philosophizing, older even than Pythagoras, Plato and Aristotle."[20] Melanchthon, in his preface to the Basel Euclid of 1537, explains that Plato insisted upon geometry as a prerequisite for entrance into his academy chiefly because its study raises our gaze, which is usually fixed on the ground, to the heavens and their Creator. Simon Grynaeus, in introducing the first printing of the Greek text (1533), likewise invokes Plato, and rates geometry as the best possible study for the "pious mind devoted to things divine." One can hardly avoid the conclusion that Dee's "Pythagorean" elucidation of 1550 was but a reworking of the traditional uplifting prolegomena to the study of Euclid.

As for the reading "physicè," that too probably did not advance beyond the ordinary. Finé, in his widely used version of Euclid's first six books, itemizes the arts and sciences based on geometry; and most later sixteenth-century editors, including Stephanus Gracilis, Commandino, Clavius, and Dee himself, copied, extended, or embellished the list.[21] The fact that no contemporary reference to Dee's lectures has been found, not even in the lengthy preface on the dignity of

19. Proclus, *Commentaires*, ed. ver Eecke (1948), pp. 16–17, 40.

20. Also in Rice, *Prefatory Epistles*, pp. 380–381. Compare the similar sentiments of Lefèvre's students Charles de Bovelles and Josse Clichtove in prefaces to mathematical books published in 1503 (Rice, *ibid.*, pp. 93, 108–109).

21. For a bibliography of the editions mentioned, see C. Thomas-Stanford, *Early Editions* (1926), nos. 6–10, 13, 18–19, 30, 41.

mathematics published by his friend Montaureus in 1551, provides further evidence that they contained little unfamiliar to the Parisians. It appears that the aging Dee, dissatisfied with the reception of his life's work and pressed to defend it, misrepresented the commendable but unexceptional Euclidean lectures of his youth as a spectacular and unprecedented achievement.[22]

Dee returned to England in 1551. He set up as a mathematical consultant with a stock that included astronomical instruments designed by Gemma, "two great globes of Gerard Mercator's making,"[23] and instructions for the use of the celestial globe, written by himself and dedicated to Edward VI. Dee had access to the studious boy king through their common mentor Cheke, who saw to it that his charge understood, and even wrote upon, the importance of astronomy.[24] Dee's offering was favorably received. Another followed, a tract on subjects intently pursued at Louvain, the sizes of planets and stars and their distances from the earth. Edward rewarded its author with an annuity of 100 crowns. Dee improved this gift in 1553 by exchanging it for two absentee rectorships, the combined income of which, some £80, was twice the salary that Cheke had had as royal Cantabridgian professor.[25]

22. Calder, *Dee*, II: 148; Montaureus, *Euclidis Elementorum liber decimus* (1551). Regarding Dee's large auditory, note that students presenting themselves for degrees at the University of Paris were required to have heard, or rather to swear that they had heard, lectures on the first six books of Euclid. Hearing Dee might have eased their consciences and their burden. Finé, *Demonstrationes* (1536), epistle dedicatory; Ross, *Studies*, p. 93.

23. *AT*, p. 5.

24. B. and H. Hellyer, *J. Br. Astr. Ass.* 82 (1972): 362–366; "The study of this art requires the greatest diligence and desire for knowledge, and sometimes a unique divine inspiration." That is not Dee but Edward, praising astronomy in a Latin exercise set him by Cheke c. 1551.

25. *AT*, p. 9; Strype, *Life of Cheke*, pp. 22–26. Like most of Dee's tracts, those dedicated to Edward were not printed, and are now lost. Often the only authority for the existence of manuscripts is Dee himself;

During the same promising years, 1551–1553, Dee served Edward's master, the domineering head of government, the Duke of Northumberland, "a man for whom no one has ever had a good word."[26] Dee tutored the Duke's children, including Robert Dudley, later Earl of Leicester and a favorite of Elizabeth's.[27] He also instructed the Duke's wife, whom he improved with tracts on the names and configurations of the constellations, and the causes of the tides. As for the greedy Duke, Dee helped him to promote a search for northern routes to the riches of the orient.[28] Dee's confidence, enthusiasm, and mathematics, not to mention his training by Gemma and Mercator, were just what was wanted to reassure uneasy investors in expensive voyages through unknown seas.

Dee favored the route to the east, above Norway and Russia, against that to the west. He may have known that Norsemen routinely sailed east as far as the White Sea, and he certainly knew that Greenland opposed a vast and uncertain barrier to the west. Whatever the ground of his advocacy, it went against the geographical opinion of his day. First, the maps of 1550, greatly exaggerating the eastern extent of Asia, placed "Cathay" about 210° east of London, and hence nearer by the west than by the east. Second, the most important contemporary maps of the polar regions offered no hope of a northeast passage. Finé's world map of

---

one may guess from some that have survived that many may have been fragments, drafts, or outlines. Their titles can nonetheless serve as a guide to his concerns. The fullest published list (which does not include the titles Dee mentioned casually) is in C. H. Cooper and T. Cooper, *Athenae cantabrigienses*, II (1861): 505–509; cf. *AT*, pp. 24–27, 74–77; Taylor, *Tudor Geography*, App. 1A; *DNB*, V: 721–729.

26. G. R. Elton, *England under the Tudors*, 3rd ed. (1969), p. 209.

27. Dee remained close to the Dudleys and their relatives the Sidneys; he later taught chemistry to Philip Sidney and his friends. P. J. French, *John Dee* (1972), pp. 126–131.

28. Elton, *England under the Tudors*, pp. 334–335; Taylor, *Tudor Geography*, pp. 89–90.

the 1530s shows an open polar sea closed (except for communication with the North Sea) by a continuous collar of land formed of Europe, Asia, and America and blocking a northern route in either direction. Mercator's mappamundi of 1538, and the Basel Ptolemy of 1561, have a large polar continent joined to Europe and Russia that reduces the Barents Sea to a narrow inlet without exit to the east.[29] These authorities, having nothing to guide them in the far north, had perforce consulted their imaginations. Dee was quite correct in regarding their guesses as no better than his own.

Northumberland's fellow investors were no doubt pleased to learn that the standard maps erred precisely where their schemes were threatened. Only a trial could decide. The promoters, later chartered as the Muscovy Company, sent Dee their pilots for instruction in cartography and navigation. The first expedition, with Dee's advisee Richard Chancellor as chief pilot, reached the White Sea in 1553; three years later Stephen Borough, another alumnus of Dee's private maritime academy, penetrated to Novaya Zemlya and found the Kara Strait. At this point the Muscovy Company, which had been granted a monopoly on explorations to the north, settled down to trade with the Russians, and abandoned the search for a northeast passage for over twenty years. Dee remained an occasional consultant.[30]

Meanwhile, he had lost his royal protectors. In 1553 Edward died; Northumberland, who tried to keep Mary from the throne, was beheaded; and several of Dee's sponsors, notably Cheke, who had been elevated to Secretary of State,

29. A. E. Nordenskiöld, *Facsimile-Atlas* (1889), pp. 89, 91, and Plates XLI, XLIII, XLV; Nordenskiöld, *Periplus* (1897), Plate XLIV; C. V. Langlois in Soc. amér. de Paris, *Journal* 15 (1923): 83–97. Cf. Taylor, *Tudor Geography*, pp. 80–81; H. Averdunk and J. Müller-Reinhard, *Gerhard Mercator* (1914), pp. 9–10, 17–20.

30. Taylor, *Tudor Geography*, pp. 89–97. Dee had at least one foreign student, the Portuguese pilot Anes Penteado; Teixera da Mota, *Mar* (1972), p. 60. Cf. A. Cortesão, *Cartografia* (1935), II: 264.

fled the country.[31] It was soon the turn of our consulting mathematician. He went to prison in 1555, on the charge of "calculing and conjuring"; but his association with Northumberland, and his impolitic advances to Princess Elizabeth, were perhaps the chief causes of his difficulty. He cleared himself by serving as chaplain and assistant inquisitor to Bishop Bonner of London, a ferocious enemy of magic.[32] Thus purged, he memorialized Mary on the need to collect manuscripts scattered or endangered by the spoliation of the Catholic foundations, a greedy, stupid thievery in which his former patron, Northumberland, had been a ringleader.[33] Mary did not support the project. Dee preserved whatever he could. He gradually built up what was probably the most extensive and important personal library of mathematical and philosophical works in England.[34]

Elizabeth's succession in 1558 again brought Dee a secure, if peripheral, connection with the Crown. His former charge, Robert Dudley, became a power in the realm; he himself performed small services for the Queen (such as "calculing" an astrologically appropriate day for her coronation) and occasionally advised or instructed her.[35] The first decade or so of her reign were Dee's most productive years. He immediately made public the "chief Crop and Roote, of ten yeres his first Outlandish and Homish studies and exercises philosophicall,"[36] namely the *Propaedeumata aphoristica*; six years later he harvested a second crop, the *Monas hieroglyphica*; and, after a further six, his "very fruitfull Preface" to Sir Henry Billingsley's translation of Euclid's *Elements*.

31. Strype, *Life of Cheke*, pp. 91–112.
32. *AT*, pp. 20, 53, 57; French, *Dee*, pp. 6–8; K. Thomas, *Religion* (1971), p. 307. The matter is thoroughly treated in Calder, *Dee*, I: 310–318, and II: 157–165.
33. Elton, *England under the Tudors*, p. 209; *AT*, pp. 46–49.
34. Yates, *Theatre* (1959), pp. 1–19; James, *List* (1921).
35. *AT*, pp. 12–23.
36. *Ibid.*, p. 56.

Among the lessons one might draw from the first book, the *Propaedeumata*, is the extreme tediousness of advancing knowledge of the physical properties of bodies by mathematical analysis. Among other impossibilities, Dee directs the conscientious astrologer to discriminate the influences of 25,000 different sorts of planetary conjunctions. As for himself, he was too impatient for such labors, and, as in the case of the traffic with Cathay, he sought a shorter path. In 1562 he wrote out a "compendious table" of the Hebrew cabala. Two years later, at the end of a second study trip abroad, he published his short way or true path, his *Monas*, "elucidated mathematically, magically, cabalistically, and anagogically." Despite these several elucidations, most of Dee's contemporaries found the book unintelligible; in contrast to the *Propaedeumata*, which was later reprinted to assist the reader of the *Monas*, the new work was accessible only to hermetists and magicians.[37] It was Dee's public declaration of a shift in his studies that had occurred in the late 1550s, and that ultimately destroyed him.[38]

The last of Dee's principal published contributions to natural philosophy were the "Preface" and supplementary theorems to Billingsley's Euclid. This, the first English version, is a beautiful folio volume, quite beyond the means (as it was beyond the grasp) of the master mechanics and advanced artisans for whom, according to Dee, it was largely intended.[39] The "Preface," an extensive reworking and elaboration of the usual prefatory themes, emphasizes the value of mathematics when applied to building, navigation, fortification, and the like. It also bursts out in hymns to the inner beauty and higher purposes of mathematics in the style of the Italian Platonists.

37. See chap. II: 4, below; T. Smith, "Vita Joannis Dee" (1707), p. 12; cf. L. Firpo, *Rinascimento* 3 (1952): 29, 33; W. I. Trattner, *J. Hist. Ideas* 25 (1964): 17–34.

38. Dee's first sustained interest in alchemy appears to date from 1556, when he borrowed books on the subject from Oxford. Corpus Christi MS 191, ff. 88–91 (Bodleian); Calder, *Dee*, II: 125, 302–305.

39. "Preface," sigs. A.iij^v–A.iiij^r.

The combination was long popular and, it appears, inspi-rational.[40] As for the supplementary material, some seventy problems, propositions, and corollaries, they are a mixture of inventiveness and puffery, of sound pedagogy and outrageous crowing.

The appearance of the supernova in Cassiopeia in 1572 provided the occasion for Dee to publish his method, probably invented twenty years earlier, for finding the parallax of circumpolar objects.[41] He did not print his observations of the nova, or his judgment of its astrological significance, and the manuscript containing them is now lost. His writings on maritime matters, which again occupied much of his time in the late 1570s, were luckier. Several characteristic ones have survived. In the most ambitious, Dee offered a plan for establishing and maintaining a British Navy; he specified its requirements of timber, men, and supplies, and a method and level of taxation for its support. In another, he squeezed dry the Biblical stories about the voyages of Solomon, decided arbitrarily on the location of Ophir, the King's gold and spice market, and advocated a voyage thither. Returning to the real world, he helped instruct the Muscovy Company's pilot, Martin Frobisher, for his first try at a northwest passage; he certified the worthless ore Frobisher brought back to be gold; and, seeing a quicker way to riches than the alchemical furnace, he invested in Frobisher's second expedition.[42] And he continued to press for another try to the northeast, once again opposing the opinions of the continental cartographers, but now on weaker grounds.

40. D. W. Waters, *Art of Navigation* (1958), p. 131; French, *Dee*, pp. 173–177; D. M. Simpkins, *Ann. Sci.* 22 (1966): 235–236; Calder, *Dee*, II: 350–351; J. E. Stephens, *Aubrey* (1972), p. 77. Cf. Samuel Butler's conjurer, who "had read Dee's Preface before / the Devil and Euclid o're and o're" (*Hudibras*, II.iii: 235–236).

41. See chap. II: 1 below.

42. Taylor, *Tudor Geography*, pp. 108–109, 114–117, 120; Calder, *Dee*, II: 370–373.

Mercator had shrunk his polar continent to four imaginary circumpolar islands. In addition, in 1569, guided by the experience of the Muscovy pilots, he had published a much-improved map of the Scandinavian and Russian coasts from North Cape east. He allowed a northeast (and, as he had for some time, a northwest) passage, and correctly placed the northernmost limit of Asia, the extremity of the promontory that closes the Kara Sea (Cape Tabin or Chelyuskin), at about 78°N. The same representation was adopted by Ortelius in his world map of 1570.[43] Dee insisted that no such promontory existed, that, after turning North Cape, one reached Cathay by a smooth sail to the southeast. He had this intelligence, "a record [he said] worthy to be printed in gold," from an excerpt from the *Geography* of Abulfeda (Abu'l-Fidā' Ismā'il), as published in the second volume of Ramusio's *Viaggi* (1559).[44] The original, composed about 1320, gives latitude and longitude for many places up to about 50°N, with an accuracy that diminishes rapidly with distance from Persia. It also describes gross geographical features, such as the lie of the Western Ocean, which "meets the northern point of the world at 71° of latitude."[45] Dee had every reason to distrust his medieval Arab authority: he knew that Greenland ran beyond 71°N, and he knew much more about the earth than his guide, who was ignorant of the existence of America and thought that Africa ended at the equator.[46] Nonetheless Dee, whose position as cartographical advisor

43. Nordenskiöld, *Facsimile-Atlas*, p. 95, and Plates XLVI, XLVII. Mercator's sketch of the lands immediately about the pole was, however, entirely fanciful, and based on no better authority than Dee invoked. Taylor, *Imago mundi* 13 (1956): 56–68.

44. Taylor, *Tudor Geography*, pp. 131–132, 137–138. In fact, the evidence Ramusio offers is his own, a globe he had seen as a boy in Venice (Ramusio, *Navagationi e viaggi*, II [1559], f. 17); it agrees with Abulfeda's.

45. Abulfeda, *Géographie*, tr. Reinaud and Guyard (1848–1883), II:1, 21–25, 31.

46. Cf. R. A. Skelton, *Duisb. Forsch.* 6 (1962), 158–170.

was passing to a new generation of less romantic geographers, grasped at the straw. The Muscovy Company's last try to the northeast (1580) probably marks the end of Dee's maritime consultancy. The expedition failed, turned back by pack ice in the Kara Sea.

No more than mathematical astrology, alchemy, or cabala did old maps and legends supply Dee with the wisdom he sought. In 1581, he hired a scryer and began to talk to angels. The following year, he wrote his last tract of any scientific importance, a plea that Elizabeth order the adoption of the Gregorian Calendar in England. His arguments from astronomy, history, and convenience did not move Anglican bishops unwilling to accept any innovation, however useful, sponsored by a Pope.[47] Meanwhile the angelic interviews intensified. They were his last hope for illumination. As he told an emissary of the Hapsburg Emperor in 1584, he had by degrees passed through "all manners of studies . . . , as many as were commonly known and more than are commonly heard of," and all ultimately unsatisfactory. "At length I perceived that onely God (and by his good Angels) could satisfie my desire, which was to understand the natures of all his creatures."[48] The attempts to extract wisdom from angels were to continue with various crystal-gazers, in England and in Europe, until the end of Dee's life. Not unreasonably, they again brought upon him the charge of conjuring. Many of his more lucid moments went to trying to clear himself of charges of invoking evil spirits and to trying to collect the pecuniary rewards that Elizabeth often promised and occasionally gave him.[49]

47. Add. MS 32092, ff. 29–31 (British Library). The bishops also appealed to the classic argument against reform, "omnis mutatio periculosa," "Mutations and alterations in Commonwealths are not to bee allowed . . . unless Necessitie inforce thereunto."
48. Dee, Relation, ed. Casaubon (1659), p. 239.
49. W. Gwyn Thomas, Welsh Hist. Review 5 (1971): 250–256.

2. DEE AS A MATHEMATICIAN

Dee was considered an accomplished colleague by contemporaries practiced in the mathematical arts. Among his countrymen, Thomas Digges (c. 1546–1595), the earliest outspoken English Copernican, esteemed him as a man "most expert in these sciences, and admirable in other philosophy." Digges's opinion has a peculiar authority, for he had been Dee's student, and was no less able a mathematician than his master.[50] Similarly Dee's friend, the well-read poet Gabriel Harvey, took him as the standard against which to measure achievement in geography and astronomy.[51] Among the less able, Dee passed as a prodigy, "a great learned man," "ye Prince of Mathematicians of this age."[52]

As for the European experts, with whom Dee had or boasted close ties, they treated him respectfully, and sometimes warmly. He is "a man of excellent wit, and singular learning," according to Federico Commandino, translator and editor of the great Greek geometers, of Apollonius, Archimedes, and Euclid.[53] Commandino knew his man personally. Dee had sought him out in Urbino, in 1563, to deliver a transcription he had made of an old Latin translation of what he guessed to be an Arabic version of a lost book of Euclid's; lacking, as usual, the energy or will to see his manuscript through the press, he picked Commandino as the best man for the task.[54] Dee's "singular library of unpublished ancient

50. Digges, *Alae seu scalae mathematicae* (London, 1573), preface, quoted by Taylor, *Tudor Geography*, p. 254; Dee, *Parall. comm.* (1573), preface; William Bourne, "Treatise on Optical Glasses" (c. 1578), quoted by Taylor, *Tudor Geography*, p. 254; F. R. Johnson, *Astronomical Thought* (1937), p. 178; Johnson and S. V. Larkey in Hunt. Libr., *Bull.*, no. 5 (1934): 105–107.

51. G. C. Moore Smith, *Harvey's Marginalia* (1913), pp. 162–163.

52. Contemporary opinions cited by Taylor, *Tudor Geography*, p. 254, and Smith, *Harvey's Marginalia*, p. 277.

53. Commandino to Francesco Maria II of Urbino, in Euclid, *Elements*, ed. Leeke and Serle, pp. 603–604.

54. *DSB*, III: 363–365; E. Rosen, *Scripta math.* 28 (1970): 321–326; P. L. Rose, *Isis* 63 (1972): 88–93.

writers of Greek mathematics" also won him the attention
of Ramus; indeed, Dee was the only British mathematician of
whom Ramus had then (1565) heard, despite diligent inquiry
among Frenchmen knowledgeable about England. "I would
be much obliged to hear from you [Ramus wrote] what old
mathematical writers are in your library, and who teaches
mathematics in your schools, and in what capacities."[55] Other
continental mathematicians who respected Dee include
Mercator, who took his cartographical observations seriously,
and Tycho Brahe, who thought him a competent armchair
astronomer.[56]

The fact of Dee's contemporary reputation is easier to
ascertain than its basis. We can dismiss the suggestion that he
was admired for "profundity."[57] He quite rightly does not
figure on van Roomen's list of the chief mathematicians of the
later sixteenth century.[58] Dee's contributions were promo-
tional and pedagogical: he advertised the uses and beauties of
mathematics, collected books and manuscripts, and assisted
in saving and circulating ancient texts; he attempted to
interest and instruct artisans, mechanics, and navigators, and
strove to ease the beginner's entry into arithmetic and geometry. It is in this last role, as pedagogue, that Dee displayed
his competence, and made his occasional small contributions
(which he classed as great and original discoveries) to the
study of mathematics.

Dee's surviving writings on technical subjects, which may
appear episodic and disjointed, take on coherence when read
as examples or extensions of the divisions of mathematics
elaborated in the preface to Billingsley's Euclid. The sources

55. Ramus to Dee, 14 Calends Jan. 1565, in Ramus, *Praefationes*
(1599), pp. 174–175.

56. Taylor, *Imago mundi* 13 (1956): 57–68; Brahe, *Astronomiae instau-*
ratae progymnasmata (Uraniborg, 1602), in *Opera omnia*, ed. J. L. E. Dreyer,
III (1916): 203–204.

57. French, *Dee*, p. 168.

58. A. van Roomen, *Ideae mathematicae pars prima* (Antwerp, 1593),
quoted by Gilbert, *Rev. quest. sci.* 16 (1884): 445.

of this justly famous piece were as wide as the Renaissance. For his main divisions of the mathematical sciences, Dee drew on Aristotle, on modern writers on algebra, and on recent chatty editors of Euclid, such as Stephanus Gracilis;[59] for his multiplication of the mathematical arts he had the example of the "incomparable architect Vitruvius," much approved by the Italian humanists, the lists of Gracilis, Recorde, and others, and the guide of his own continental studies.[60] As regards the sublimer side of the subject, Dee invoked Plato, Boethius, and latter-day epigones like Pico della Mirandola; on the seamier side, on mathematical magic, mechanical contrivances, games and spectacles, he called up Moses, Zoroaster, and the odd abbot Trithemius.

Arithmetic

Arithmetic, the study of number immaterial, is the highest form of mathematics. Number, as we have seen, was the exemplar or pattern in the mind of the Creator; "all creatures distinct virtues, natures, properties and *Formes*" are comprehended by it. Dee promises that a knowledge of these arcana will prepare us to see the "*Forme of Formes*, the *Exemplar Number* of all things *Numerable*." If lucky, we may find "the number of our own name, gloriously exemplified and registered in the booke of the *Trinitie*," the very number that the Creator continually counts to conserve us in our being and state.[61]

59. Dee's copy of Gracilis's *Euclidis elementorum libri XV* (Paris, 1557), bought in 1558, is in the British Library. Gracilis sets out the two divisions of mathematics transmitted by Proclus, the Pythagorean (arithmetic, music; geometry, astronomy) and Geminus's (arithmetic, geometry; astrology, optics, music, practical arithmetic, geodesy, mechanics), and lists several applied branches. Proclus, *Commentaires*, pp. 29–33.

60. E. Kaplan, *Robert Recorde* (1960), pp. 67–70; Yates, *Theatre*, pp. 21, 27, 33, exaggerates in making Vitruvius the chief inspiration for the "Preface."

61. "Preface," sigs. *j^r–*ij^r.

But there are cruder uses of number, namely, the practices of merchants, lawyers, and quartermasters, practices so
impure as to require fractions and, worse, fractions of fractions,
square roots, cube roots, irrationals of all kinds. Dee did not
fear these nettles. In 1561, and again in 1570, he prepared new
editions of Robert Recorde's *Ground of Arts*, an introduction
to arithmetic first published in 1543. Recorde (c. 1510–1558),
a Cambridge graduate and an Oxford M.D., shared many of
Dee's ideas, to which, however, he gave different emphases.
He accepted, but did not develop, the principle that "God
was alwaies workinge by Geometrie"; his business was to
exhibit the elementary operations of mathematics plainly and
briefly, and to illustrate their application to the affairs of
practical men.[62] His books were extremely popular: the
Muscovy Company sponsored some of them as manuals for
their pilots, and it was perhaps under the Company's com-
mission that Dee bestirred himself to enrich the *Ground*
shortly after Recorde died in King's Bench Prison. It appears
that Recorde was imprisoned in consequence of irregularities
or incompetence in his administration of the mines and
moneys of Ireland, of which he became "surveyor" in 1551;[63]
the application of mathematics is more easily preached than
practiced. Dee's improvements of the *Ground of Arts*, though
slight enough,[64] had the merit of bringing him to the business
of adult education, and to the obligation of explaining himself
clearly. His collaboration with Billingsley was perhaps guided
and inspired by his service to Recorde.

62. Recorde, *Pathway* (1551), preface; cf. Kaplan, *Recorde*, pp. 112–
115; *DSB*, XI: 338–340; J. B. Easton, *Scripta math.* 27 (1966): 339, 346.
Recorde's innovations in pedagogy are discussed by Johnson and Larkey
in Hunt. Libr., *Bull.* 7 (1935): 59–87.

63. F. M. Clarke, *Isis* 8 (1926): 50–70; Kaplan, *Recorde*, pp. 36–37.

64. Primarily extension of the sections on fractions and addition of
problems and tables of foreign exchange and weights and measures;
Easton, *Isis* 58 (1967): 529–531. A fragment on fractions by Dee, which
concerns multiplication, division, reduction to least common denomi-
nator, and mixed numbers, is in the Bodleian, Ashmole 242, ff. 160v–156v
(it is bound upside down).

The highest form of mundane arithmetic is "algiebar," "so profound, so generall," according to Dee, that "man's wit can deale with nothing more profitable about numbers."[65] He liked the subject, which was perhaps the liveliest mathematics of his time; and, as will appear, he preferred those parts of geometry, such as the theory of proportion, that lent themselves to algebraic manipulation. The "Preface" contains an interesting application of algebra—or "the great art" as Dee's acquaintance Girolamo Cardano called it[66]—to a standard problem in sixteenth-century medicine: to determine the resultant temperament of a compound of drugs each hot (or cold) and moist (or dry) in a given degree. Many solutions had been proposed, some of great complexity; Dee ignored the chief difficulty, the physics of mixing, in favor of a quantitative description that he found sketched out in an old manuscript of Roger Bacon.[67]

In Bacon's arbitrary scale, quality can have any degree from zero to four, and opposites coincide at zero. Consider the mixing of equal quantities of two medicines, one moist in the first degree and the other dry in the third. The resultant is dry in the first: the middle quality is $(3 + 1)/2 = 2$, and the resultant is therefore two degrees from the more intense toward the lesser, or at $3 - 2 = 1$ degree dry. From this and other numerical examples Dee gives, one may deduce a general rule, $r = a - (a \pm b)/2$, where r is the resultant and $a \geq b$ the initial degrees. The plus sign is to be taken when a

65. "Preface," sig. *ij^v. Gemma and Nuñez were also enthusiastic students of algebra. H. Bosmans in Brussels, Soc. scient., *Annales* 30:1 (1906): 165–168, and *Bibl. math.* 3 (1908); 154–169. Cf. Gomes Teixeira, *História*, 167–183.

66. Cardano, *Ars magna* (Nuremberg, 1545). Dee had met Cardano in London in the early 1550s; they amused themselves by pouring vinegar on a stone, apparently a soft carbonate, which hopped about as it released carbon dioxide. R. R. Steele, *Notes & Queries* 1 (1892): 126; Cardano, *Book of My Life*, tr. Stoner (1962), pp. 97–100.

67. "Preface," sigs. *iij^r–*iiij^v; N. H. Clulee, *Ambix* 18 (1971): 178–211.

and b are opposite (hot and cold, or moist and dry), the negative when they are alike. A more interesting case, and more algebra, arises when the quantities of the components differ. Dee represents the "coss," or unknown resultant, by x;[68] the quantities he takes as integral multiples of an arbitrary unit m. His general rule and the examples given amount to the formula $p:q = (a - x):(x - b)$, where $a \geq b$ and pm, qm are the quantities of the ingredients at degrees a, b, respectively. In his discussion, Dee makes use of the equal sign, introduced by Recorde,[69] and implicitly admits that degrees of cold (or dry) are algebraically the negative of degrees of hot (or moist). He does not, of course, state explicitly the strict algebraic equations just given. No sixteenth-century mathematician would have done so.

Like much of Dee's work, the little analysis of the graduation of qualities is not profound or original, but it points in a fruitful direction. It can, in fact, be tested and confirmed. If x, a, and b stand for temperatures, Dee's expression becomes the calorimetric mixing formula established in the eighteenth century.[70] Application of this formula resulted almost immediately in the capital discoveries of latent and specific heats and the beginning of a quantitative science of thermodynamics. Since the eighteenth century the reach of mathematics in the description of physical phenomena has steadily increased, and is today so extensive and successful that it seems a matter of course. To Dee it was (as it is) deserving of the greatest admiration. Let us praise God, he said, and marvel, "that the profoundest and subtilist point, concerning the *Mixture of Formes and Qualities Naturall*, is so Matcht and

68. "Coss" is the German and English form of Italian "cosa," thing: Dee uses the notation of Stifel.

69. F. Cajori, *History*, I (1928): 165–166, 168.

70. V. P. Zubov, in Akad. nauk, Inst. ist. estest. tekh., *Trudy* 5 (1955): 69–93, and in *Mélanges A. Koyré* (1964), I: 654–661; D. McKie and N. H. de V. Heathcote, *Discovery* (1935).

maryed with the most simple, easie, and short way of the noble Rule of Algiebar."[71]

Geometry

The highest branch of geometry is the study of pure magnitude; Dee proposes to call it "megathology," to free it from the earthy implications of "geometry." By "Megathologicall Contemplations" we can rise almost as high as we can by arithmetic. We can train our minds gradually to "foresake and abandon the grosse and corruptible objects of our outward sense, and to apprehend, by some doctrine demonstrative, Things Mathematicall"; and from there we can ascend to "conceive, discourse, and conclude of things Intellectual, Spirituall, aeternall, and such as concerne our Blisse everlasting."[72] As helps to those just starting to climb—"to geve to young beginners some light, ayde and courage"[73]— Dee improved a few of the usual Euclidean diagrams, supplied an occasional omission, and added "certaine most profitable Corollaries, Annotations, Theoremes and Problemes" to the English Euclid.[74]

Most of these items, which are scattered through the last three books of the *Elements*, concern the finding of a line or

---

71. "Preface," sig. *iiij^v.

72. *Ibid.*, sig. a.iij^r.

73. Euclid, *Elements*, ed. Billingsley (1570), f. 362^r.

74. *Ibid.*, f. 346^r. Billingsley credits Dee with repairing the diagrams for Props. XI.29, XI.31, and XIII.17 (ff. 337^r, 340^v, 377^r), emendations that have become standard (T. L. Heath, *Thirteen Books*, 2nd ed. (1925), III: 333, 337). As for omissions, Dee improved the faulty enunciation of XI.24 (Billingsley, f. 337^r; Heath, III: 324), and supplied a uniqueness proof after XIII.2 (Billingsley, f. 391), showing that a given line can be divided into mean and extreme proportion in only one way. Dee's contributions will be found on ff. 255^v–256^r, 268, 275, 286^v, 306^r, 311^r (Book X); 325^v–326, 329^v, 337^r, 340^v, 342^r, 346–347, 348^v–349, 352 (Book XI); 356^v–357, 359–362, 371, 376^r, 379^r (mispaginated as 372), 380–389 (Book XII); 391–394^r, 395, 397^v (Book XIII). Cf. *AT*, p. 73, and R. C. Archibald, *Amer. Math. Monthly* 57 (1950): 443–450.

lines in a set ratio to given lines, areas, or volumes. For example, in a corollary after XI.33,[75] Dee shows how to find two lines whose ratio is that of the volumes of given cubes. One first constructs an intermediate line, y, say, as a third proportional to the sides of the cubes a and b ($a:b::b:y$), and then obtains a third proportional (z, say) to b and y ($b:y::y:z$). Then $a:z::a^3:b^3$. Similarly (additions to XII.2) one can construct two squares proportional to two given lines and equal in area to a given square. In Figure 1, AB, the diameter of the

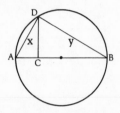

FIG. 1.   Construction for Dee's additions to Euclid XII.2.

circle, is the side of the given square s. Divide AB at C in the ratio $p:q$ (AC:CB::$p:q$); then AD = x and DB = y are the sides of the squares sought. For, from Euclid's discussion of proportion, $p:x::x:s$ and $q:y::y:s$, whence $x^2:y^2::p:q$; and ADB is a right angle, whence $s^2 = x^2 + y^2$. This construction can be extended to any number of squares $x, y, z, \ldots$ For three, we wish $x^2:y^2::y^2:z^2::p:q$ and $s^2 = x^2 + y^2 + z^2$. Divide s into three segments, x_1, y_1, z_1, in continued proportion to p and q ($x_1:y_1::y_1:z_1::p:q$). By construction, $(x_1 + y_1 + z_1)s = s^2$. Then x, y, z can be obtained as the mean proportionals between x_1 and s, y_1 and s, and z_1 and s ($x_1:x::x:s$, etc.).[76]

To spirit up the ascending Megathologian, Dee advertises these intricacies as "the key of one of the chiefe treasure houses, belonging to the state Mathematicall," as the clue to "infinite . . . , strange and incredible speculations and

75. *Euclid*, ed. Billingsley, f. 346.
76. *Ibid.*, ff. 360^r–361^v.

practices."[77] He excuses himself for not offering an example of these sublime applications on the ground that he "utterly [lacks] leisure"; he must run to similar exercises with cones, spheres, cylinders, and spherical sectors, lifting the requisite propositions from Archimedes. One exercise will suffice:[78] to cut from a sphere two sections the surfaces of which shall stand in a given ratio. From Archimedes, the surface of the spherical section XZY is equal to the area of a circle of radius XZ (Figure 2). Dee's construction (Figure 3): cut the diameter

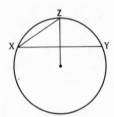

FIG. 2. Archimedes' proposition on the area of spherical sections: the area of the section XZY is πXZ^2.

of the given sphere at D into two segments in the given ratio; raise the perpendicular at D until it intersects the great circle AEFC at B; then AEB and BFC are the sections sought. (The

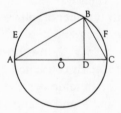

FIG. 3. Dee's application of Archimedes' proposition to the construction of spherical sections the areas of which stand in a given proportion.

77. *Ibid.*, f. 362^r.
78. It is one of 12 problems and 8 theorems added after XII.18 (*ibid.*, ff. 386^v–387^r).

areas of the sections are as $AB^2 : BC^2$, which, from Euclid XIII.13, are as $AD:DC$.)

There are two noteworthy points about Dee's problems. First, they are inspired by the ancient, unresolved problems of the squaring of the circle and, above all, the doubling of the cube. For example, as a corollary to his problem on XI.33, Dee asks for the sides of cubes the volumes of which stand in a given ratio, say as p to q; and he demonstrates the "famous lemma" (of which Finé had made extensive and disastrous use) that the sides sought, x and y, are two mean proportionals between p and q ($p:x::x:y::y:q$). Evidently, if $p:q::2:1$, the cube x is double the cube y. Dee knows that no one had yet found a way to construct such proportionals: "whosoever shall achieve that feate [he says] shall not be counted a second *Archimedes*, but rather a peerless Mathematician."[79] He did not himself despair of occupying such a height. In a problem added to XI.36, he shows how to find lines x, y, z such that $x:y::y:z::a:b$ and $xyz = c^3$, where a, b, c are given. (He takes $z = c$, $x:c::a:b$, $c:y::a:b$.) "Listen to this and devise, you couragious Mathematicians: consider how nere this creepeth to the famous Probleme of doubling the Cube." Of course the proposition is useless. "I leave as now, with thys marke here set up to shoote at. Hit it who can."[80]

The second point about Dee's problems is that, despite their attachment to ancient conundrums, they represent a progressive current in sixteenth-century mathematics: although he usually sets up his problems and manipulates his proportions geometrically, his treatment is strongly algebraic in spirit. The examples so far given show his tendency to set up equations (as proportions) and to juggle them until a solution emerges in the form of a constructible line. Another example is his apparently uninspired gloss on X.32, which rings changes on the relations $AC \cdot AB = AD^2$ and $BC \cdot AB =$

79. *Ibid.*, f. 346; Ross, *Studies*, pp. 225–235, 243–258.
80. *Euclid*, ed. Billingsley, f. 352; cf. f. 349^r.

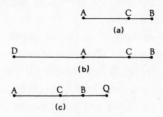

FIG. 4. Lines cut in mean and extreme (or golden) ratio: (a) the defining cut, where AB:AC::AC:CB; (b) a derived case, DA=AB, whence DC is cut in golden ratio at A; (c) a second derived case, AC=CQ, whence CQ is cut in golden ratio at B.

DB² (Figure 1).⁸¹ Again, consider his treatment of mean and extreme section. Euclid XIII.3 shows that if a line AB (Figure 4a) is so divided at C (i.e., if AB:AC::AC:CB), then the square with side CB + AC/2 is five times the square with side AC/2. Dee proves the converse proposition, and uses it to demonstrate that, if BA is extended an equal distance to D (DA = AB, Figure 4b), or if AB is extended to Q such that AC = CQ (Figure 4c), then DC is cut in mean and extreme section at A, and CQ is likewise cut at B.⁸²

In these demonstrations, Dee operates directly with proportions; he does not display the squares in question and add them up as in Euclid XIII.1–5. Later, in corollaries to XIII.5, Dee suggests alternative proofs that appeal still less to geometric intuition. He observes that by standard transformations of the defining condition AB:AC::AC:CB, one can easily show that DC (Figure 4b) and CQ (Figure 4c) are cut in mean and extreme section. For, in the first case, (AB + AC):AC:: (AC + CB):CB, or DC:AC::AB:CB; by alternation DC:AB:: AC:CB; by definition, AC:CB::AB:AC; whence DC:AB:: AB:AC, or, what is the same thing, DC:DA::DA:AC. In the second case, (AB − AC):AC::(AC − CB):CB, or CB:AC::

81. *Ibid.*, ff. 255ᵛ–256ʳ.
82. *Ibid.*, ff. 392ʳ–393ᵛ.

BQ:CB; by inversion, AC:CB::CB:BQ, or, what is the same, CQ:CB::CB:BQ.[83]

Again, none of this is work of much originality. But neither is it orthodoxly Euclidean; it looks beyond the *Elements* toward a unification of geometry and arithmetic. Dee understood that well enough. "My desire is somewhat to furnish you, toward a more general art Mathematical then *Euclides Elements*, (remayning in the terms in which they are written), can sufficiently helpe you unto." He also understood that he had not got very far, that his amplified Euclid did not run "in one Methodicall race towards any marke."[84]

Applications

The first of the geometrical arts is that of the "Mechanicien" or "Mechanicall workman," who makes sensible objects as faithful as possible to mathematical forms. Since he cannot work malevolent matter with absolute precision, he must content himself with approximations, such as the Archimedean ratio between the circumference of a circle C and its diameter d ($C:d::22:7$). By making this or another rational numerical approximation to π, the mechanician, "the vulgar and Mechanicall workman," wins a power beyond the reach of even the cleverest megathologian. He can square a circle. He need only divide a diameter into 14 parts, and make a rectangle with sides d and $(11/14)d$; the rectangle, of which the area A approximates that of a circle ($A = (11/14)d^2 \simeq \pi (d/2)^2$), can then be made into a square by the procedure of Euclid II.14. "The great Mechanicall use [this and similar operations] may have in wheeles of Milles, Clockes, Cranes, and other engines for water workes, and for warres, and many other purposes, the earnest and wittie Mechanicien will soon boult out, & gladly practice."[85]

83. *Ibid.*, f. 395.
84. *Euclid*, ed. Billingsley, f. 371.
85. *Ibid.*, f. 357; "Preface," sig. a.iiij[v].

The mechanic can double the cube as easily as he can square the circle. Let him make an open cubical box of side a, and a capacious cone; and let him mark off the slant height of the cone in 60 or 100 equal parts. He fills the cube with water, pours it off into the inverted cone, and notes the slant height of the water, p; he repeats the operation, and records the new height, q. Then the side of the cube sought, x, may be found by the golden rule ($a:x::p:q$). This is not a proper solution to the ancient problem, however, because it requires a graduated scale; but it might appear a splendid achievement to readers not alert to Dee's oscillation between the mathematical and the mechanical mode. He takes full advantage of the possibilities for equivocation. In the body of the *Elements*, he intimates that he can double the cube mathematically. In the "Preface," after giving his mechanical solution, he celebrates himself as if he had made good his promise. "For this, may I (with joy) say *EYPHKA, EYPHKA, EYPHKA*."[86]

Next come the true applications of geometry, and first "geometrie vulgar," surveying, mensuration, gauging, geography, chorography (mapping of localities), hydrography, and "stratarithmetrie," the science of military formations. Dee's contributions to these subjects included idiosyncratic maps, such as a representation in 1580 of Elizabeth's geographical rights, which, on the basis of Arthurian legends, Dee made to run from Florida to Novaya Zemlya.[87] There is also the equidistant polar projection of c. 1582 that shows Cape Tabin to terminate south of North Cape. It was perhaps the last of a series of Dee's maps used by English pilots in voyages to the north, where the rapid convergence of the meridians makes the common rectangular plane chart useless.[88] Nor is it a simple business to make the equidistant polar projection an effective navigational instrument. To assist his pilots, Dee invented in the 1550s a "paradoxall compass,"

86. "Preface," sigs. c.ij^r–c.iij^v.

87. Reproductions of these maps are given by Taylor, *Imago mundi* 12 (1955): 103–106, and *ibid.* 13 (1956), opp. p. 56, respectively.

88. Cf. Waters, *Art of Navigation*, pp. 209–211.

"whereof," as he said in the "Preface," no doubt correctly, "scarcely foure, in England, have right knowledge."[89]

The secret of the business is that a rhumb line—a course of fixed compass bearing—does not, in general, coincide with a great circle on the surface of the globe. Consider a course to the NE commencing at the equator, and ignore the problem of magnetic variation.[90] The path of the ship, which must cut all meridians at 45°, will be a species of spiral: as the meridians converge toward the north pole, the length of a degree of longitude shrinks; in raising one degree of latitude, which remains constant in length, increasingly more degrees of longitude are traversed. If one wishes to reach 45°E, 45°N from 0°0° on a constant rhumb, one must set the compass N of NE, and the path so described will intersect the arc of the great circle connecting the points. According to Dee's calculations, a ship sailing NE from the equator covers 50°30' in longitude before it reaches a latitude of 45°; one pursuing the seventh rhumb, at an angle of 78°45' to the meridian, travels almost 700° in longitude in climbing from the equator to latitude 80°, the highest to which Dee carried his calculations. In the last case, the ship's path, or "loxodrome," takes it almost twice around the earth. Evidently it is neither economical to sail at a fixed bearing over long distances, nor easy to know what other course to pursue between widely separated points.

The problem of loxodromic navigation came to light in the 1530s. Pedro Nuñez, for once not entirely accurate, obtained an approximation to their geometry in 1537. Mercator, alerted to the problem by Nuñez and Gemma, drew correct spiral loxodromes on his globe of 1541.[91] Dee learned about these problems in Louvain. During his early years as advisor to the Muscovy Company, he devised an

89. "Preface," sig. a.iiij^v; AT, p. 26; Taylor, Tudor Geography, p. 263.

90. The fact that the compass does not, in general, point to geographic north.

91. J. Keuning, Imago mundi 12 (1955): 16–18; Gomes Teixeira, História, pp. 109–122.

instrument and calculated tables that may have enabled a trained navigator to plot a course approximating to a great circle. Either the instrument, or, more likely, charts with the spiral loxodromes represented, constituted the "paradoxall compass."[92] Unfortunately, Dee's instructions for its use have not survived. We have, however, the calculations already referred to, which give, for each of the seven rhumbs, the easting or westing (the amount of longitude traversed) in raising a single degree of latitude.[93] With their help, one could draw the loxodrome corresponding to any contemplated course directly upon a globe. The error of the projected course would then be apparent, and appropriate corrections could be made.

What was wanted was a chart from which the fixed bearing between widely separated places could be read directly. On the standard plane chart, on which all parallels have the same length, the rhumbs are once again spirals, as appears from OABC in Figure 5. If, however, the chart is

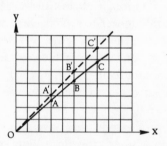

FIG. 5.   A portion of a loxodrome OABC on a plane chart; Mercator's projection amounts to stretching the chart northward so that A comes to A', B to B', C to C'.

92. Taylor in *Journal of the Inst. of Navigation* 8 (1955): 318–325, concludes that the paradoxall compass was a sailing chart on a polar projection on which the rhumbs, which pilots supposed to be straight, "paradoxically" (i.e., against the orthodox teaching) appear as spirals.

93. Dee's table, "Canon gubernauticus," has been published in W. Bourne, *Regiment*, ed. Taylor (1963), pp. 415–433; cf. Taylor, *Tudor Geography*, p. 264, and Waters, *Art of Navigation*, pp. 525–526.

stretched toward the north in such a way that A rises to A',
B to B', etc., the loxodromes become straight lines, and one
can immediately read off the bearings. That is precisely what
Mercator accomplished with the projection that he introduced
in 1569. Its utility was not immediately understood, partly
because of the complexity of the geometry, which Mercator
did not explain,[94] partly because of the great distortion of the
northern lands. It appears that Dee did not understand
Mercator's cartographic principles any better than Frobisher
understood his.[95]

After vulgar geometry, and declining further from the
"purity, simplicitie, and Immateriality, of our Principall
Science of *Magnitudes*," come the "Mathematicall Artes":
Perspective, Astronomie, Musike, Cosmographie, Astrologie,
Statike, Anthropographie, Trochilike, Helicosophie, Pneu-
matithmie, Menadrie, Hypogeiodie, Hydragogie, Horometrie,
Zographie, Architecture, Navigation, Thaumaturgike, and
Archemastrie.[96] Dee's activity in these jabberwocky dis-
ciplines centered on the first five, Perspective through
Astrologie, from which he gathered the materials for *Pro-
paedeumata aphoristica*. We shall examine them in their place.

94. Let the point x,y of the plane chart correspond to the point of
latitude λ and longitude ϕ on the globe. The scale of the plane chart has
been magnified east-west (parallel to the axis of x) by a factor sec λ
$(x(\lambda,\phi) = \phi \sec \lambda)$; along the meridians, which run parallel to the y-axis,
$y = \lambda$; along x, therefore, $dx = d\phi \sec \lambda$, and along y, $dy = d\lambda$. If one
wishes to preserve directions so that a line making a given angle with the
axis of y at the equator ($y = 0$) does so at all other latitudes, the scale of y
must be expanded to match that of x: $dy \rightarrow d\lambda \sec \lambda$. This amounts to
pulling the plane chart of Figure 5 north in such a way that the point
formerly at $y = \lambda$ is now at $y' = \int_0^\lambda d\lambda \sec \lambda$. It appears that Mercator,
who did not have the mathematics to approximate these integrals, drew
his map after projecting the loxodromes of his globe onto a plane chart.
B. Kyewski, *Duisb. Forsch.* 6 (1962): 115–130.

95. Taylor, *Imago mundi* 12 (1955): 103–104; Frobisher and Hall to
Dee, 26 June 1576, in Taylor, *Tudor Geography*, p. 262

96. A list still endorsed in the seventeenth century; see Webster,
Academiarum examen, p. 52.

To several of the other sciences Dee contributed tracts or treatises now lost: to Trochilike, "which demonstrateth the properties of all Circular Motions," very useful to millwrights, wheelwrights, and all who exploit the power of water wheels;[97] to Hypogeiodie, the art of straight tunneling and underground surveying;[98] to Zographie, "the Scholemaster of Picture, and chief governor," the almost divine art of correct rendering in painting;[99] and to Navigation, the best way to bring a ship between any two points assigned, an art requiring knowledge of many geometrical sciences as well as of tides, of coasts, and of the habits of birds.[100]

To Statike, Dee donated an endorsement of the forward-looking ideas of G. B. Benedetti on the nature of the heavy and the light, ideas that later, and more fruitfully, inspired Galileo;[101] and a ringing prayer for revelation of the third key to creation, the weights that went with the original numbers and measures. Following Vitruvius, Dee made Architecture the "storehouse of all workmanship," a "naturall subject of . . . the pereless *Princesse Mathematica*"; an advertisement decisive, it is said, for the acceptance of Renaissance, or neo-Vitruvian, ideas about architecture in Elizabethan England.[102] Horometrie, the science of time-keeping, profited—or would have, had the bishops not had their way—from Dee's useful account of the Gregorian calendar. As for

97. "Preface," sig. c.iiij^v; "Trochilica inventa mea" (MS of 1558), *AT*, p. 26.

98. "Preface," sig. d.j^v; "De itinere subterraneo" (MS of 1560), *AT*, p. 26.

99. "Preface," sig. d.ij^v; "De perspectiva illa, qua peritissimi utuntur pictores" (MS of 1557), *AT*, p. 26.

100. "Preface," sigs. d.iiii^v–A.i^r; "The British Complement of the Perfect Art of Navigation" (MS of 1576), *AT*, p. 25.

101. "Preface," sigs. b.iij^r–c.j^r. Benedetti taught that all bodies made of the same material fall at the same rate, irrespective of size. Cf. A. Koyré, *Études galiléenes* (1939), I: 41–53, and Galileo, *On Motion and Mechanics*, ed. Drabkin and Drake (1960), pp. 28–30.

102. "Preface," sig. d.iiij^r; Yates, *Theatre*, pp. 20–41, 190–197.

the arts Anthropographie (an astronomy of the lesser world, the microcosm, the "perfect body of man"), Helicosophie and Menadrie (the study of simple machines), Pneumatithie (the application of nature's abhorrence of emptiness to siphons, bellows, and pumps), and Hydragogie (hydraulics), Dee enriched them only with Greek names now happily forgotten. The same, alas, is true of Archemastrie, the science of science, *scientia experimentalis*, the way to the "chief and finall power of Naturall and Mathematicall Artes."[103]

There remains Thaumaturgike, the art of making "straunge workes, of the sense to be perceived, and of men greatly to be wondered at." Dee has in mind the old tricks of Ctesibius and Hero of Alexandria, Archytas's dove, and automata like the iron fly of Nuremburg.[104] He had tried his hand at such magic early in his career, with calamitous results. As Fellow of Trinity he assisted in a presentation of Aristophanes' *Pax*, in which a great beetle ascends to Jupiter's palace with a man on its back. Dee's device for effecting this rise made a great impression. He later liked to think that it was the cause of his reputation for conjuring, which he could therefore make out as baseless, as the "folly of Idiotes and the Mallice of the Scornfull."[105] And so one finds at the end of his "Preface" in praise of mathematics an indignant defense of self, an incongruous, emphatic denial of the charge of trafficking with the devil. Should he, he asks, a modest Christian philosopher, after twenty or twenty-five years of study and great expenditures,

103. "Preface," sig. A.iij.

104. Proclus, *Commentaires*, p. 35, classes the ancient tricks as applied mathematics. For the iron fly, which buzzed about and returned to its master when tired, see G. Naudé, *Apologie* (1712), p. 51, and W. Shumaker, *Thought* 51 (September 1976): 255–270.

105. *AT*, pp. 5–6; "Preface," sig. A.j^v; cf. Yates, *Theatre*, 30–32, and Calder, *Dee*, I: 194, 280–281. "Les plus excellens Mathematiciens ont toujours été soupçonnez de Magie," says Naudé, pointing to Pope Sylvester, Roger Bacon, Michael Scot, Albert the Great, and "ce Jean Denys [!] excellent Mathematicien de nôtre temps, qui fit imprimer une apologie pour sa defence, l'an 1570."

after fearful travels for learning's sake, "in all manner of weathers, in all manner of waies and passages, both early and late, in danger of violence by man, in danger of destruction by wilde beastes," should he, after such travail, bargain with a fiend, a hypocritical crocodile, an envious basilisk? "Should I (I pray you) for all this ... have fished with so large and costly a Nette, so long time in drawing (and that with the helpe and advise of Lady Philosophie, & Queene Theologie): but at length, to have catched, and drawn up a Frog? Nay, a Devill?"[106]

Was the charge of the "common peevish Pratler" right? Did Dee's philosophical fishing net him only a frog? Or, as some historians believe, did he catch pearls of great price, the first jewels in the treasury of Galileo, Descartes, and Newton?

3. DEE AND THE SCIENTIFIC REVOLUTION

There is no doubt that Dee was a competent and knowledgeable mathematician and that he played an important part in making the mathematics and the occultism of the Continental Renaissance known in England. His technical advice and lessons and his constant advertisements for mathematics, culminating in the stirring long-winded "Preface," encouraged people to study geometry who might otherwise have neglected it. Several of the mathematical masters active in London at the end of the sixteenth century learned their art either directly from him or from others inspired by him. All this gives Dee a modest place in the intellectual history of Tudor England. To this may be added a part in justifying,

---

106. "Preface," sig. A.ij^r. As one would expect, this is a frequent theme with Dee; cf. *AT*, pp. 4, 55–57, 62, 71–72, 79, and Dee to Lord Burghley, 30 Oct. 1574 (Taylor, *Tudor Geography*, p. 259): "I have had a marveillous zeal, taken very great care, endured great travail and toil, both of mind and body, and have spent many hundred pounds, only for the attayning some good and certain knowledge in the best and rarest matters Mathematicall and Philosophicall."

encouraging, and attempting to realize the imperial ambitions of Elizabeth.[107]

Recently historians have found a larger role for Dee: he was not only an important link between English and Continental savants, but also a prime mover and necessary forerunner of the Scientific Revolution. Here is the opinion of Peter French, the latest Dee monographer: "The more overtly scientific attitudes of men like Francis Bacon and William Harvey evolved from the approach of Dee and his [magician] colleagues. . . . Dee's theories about mathematics, architecture, navigation and technology—all part of a broader magically oriented philosophy—achieved results: they helped to pave the way for the momentous scientific advances of the seventeenth century."[108] Dr. French takes his theme from Frances Yates of the Warburg Institute. Dr. Yates, an imaginative and resourceful historian, has come to see the Renaissance magus as "the immediate ancestor of the seventeenth-century scientist," and the magicians' frame of mind as "the necessary preliminary to the rise of science."[109] Her argument runs roughly as follows.

The science of Galileo, whom no one refuses a place in the revolutionary van, is characterized above all by the application of mathematics to physics and by appeal to experiment; it also disdains the school philosophy, finds some of its concerns in the technical problems of engineers, mariners, and military men, and uses special materials, instruments, and apparatus. According to Yates, the same characteristics may be found in Renaissance hermeticism, which she makes a mixture of Neoplatonism, magic, numerology, cabala, alchemy, and the

107. Dee was not the CIA man that Richard Deacon portrays in his worthless fiction, *John Dee, Scientist, Geographer, Astrologer, and Secret Agent to Elizabeth I* (London, 1968). Deacon has all Dee's gullibility, but none of his learning.

108. French, *Dee*, pp. 19, 187.

109. F. A. Yates in *Art, Science, and History*, ed. Singleton (1968), pp. 255, 258.

teachings of Hermes Trismegistus. Her mixture may be dissolved into two distinct ingredients.

First is the Neoplatonic/magical, which rests upon the old scheme of Emanations from the original, indifferent One. The Emanations, ordered in strength and dignity, each in turn created and creating, decline from the first immaterial hypostases down to and beyond the first visible incarnation, the sun, the material image of God. The hermetist believed himself capable of tapping the energies, or exploiting the "sympathies," in the chain of creation that holds the cosmos together. Alchemy and astrology hinted how he might proceed, how the sun's creative power or the derivative planetary influences might be concentrated in appropriate talismans, amulets, and symbols. In performing these operations, in bringing together the magical material, in resorting to his furnace, burning mirror, harp, or dung heap, the hermetist became a magus and, according to Yates, the ancestor of the experimental physicist.[110]

The second ingredient in Yates's hermeticism is Pythagorean/cabalistic. It links to the hermetic complex by locating Number among the early Emanations, among the Ideas sprung from the One, and by invoking practical cabala, or number magic, to help tap the Neoplatonic power line. And it links or leads to Galileo in two ways: it endorses the promiscuous study and use of mathematics, including the applications of arithmetic and geometry spelled out by Dee; and it authorizes the natural philosopher to search for the simple mathematical relations among physical quantities that constitute the Pattern of the Universe.

Yates's thesis gains much of its plausibility from its inclusiveness. There is scarcely an interesting writer on mathematics or physics from Ficino to Galileo who does not touch one or another of her hermetic themes. But that does not make them magicians. We know, for example, that from the first Parisian

110. Cf. *ibid.*, p. 272.

printing of Euclid in 1516 the *Elements* customarily carried a preface praising Plato; yet few editors or students were Pythagoreans, much less hermetists. One can even be a magician and not a hermetist. Whoever taps the world's energies or sympathies merely by manipulating ordinary, given powers, practices natural magic, Dee's thaumaturgy; he designs mechanical tricks, makes up philtres and cosmetics, invents cryptograms, plays with mirrors and magic lanterns, in short does the thousand amusing, silly, practical, or useless things set forth in the books of Giambattista della Porta, Athanasius Kircher, Gaspar Schott, and Francesco Lana. If, on the other hand, one attempts to influence the sources of power, to alter the course of nature, one must cajole alien intelligences; one sings them songs or shows them pictures, summons spirits, invokes demons, conjures up angels good and bad. These were the tendencies of true hermetic magic. The natural variety could be respectable and successful; its practitioners acquired knowledge and experience of materials, particularly metals and glass, that went into building the apparatus of the scientific revolution. Needless to say, invoking spirits was useless and dangerous; conjurers might learn at great risk to their souls the hierarchies of demons and the unutterable names of God, but their exuberant animism blocked investigation of the ordinary and the regular, and they ended no wiser than they began.

Some necessary distinctions

In the 1920s Lynn Thorndike began to uncover what he took to be examples of the intercourse and offspring of magic and empiricism, of alchemy, astrology, mysticism, medicine, and natural history. From Thorndike's huge and heterogeneous inventory, Paolo Rossi has chosen Francis Bacon as a capital test case. Certainly Bacon, whom many seventeenth-century scientific revolutionaries took as their warlord, profited from the writings of the magicians. He took the magus to be a manipulator, a midwife, able to accelerate, retard, or prepare the operations of nature; an impresario who thoroughly understood natural powers and sympathies and, what is more, the terminology in which sixteenth-century adepts expressed them. Bacon saw in the industrious study and subsequent manipulation of natural powers the needed antidote to his favorite evil, the ills of traditional learning: "the aim of magic is to recall natural philosophy from the vanity of speculations to the importance of experiments."[111]

But is Bacon's magic that of the adepts of Hermes? Rossi observes that although Bacon accepted some of the metaphysical principles of hermeticism, such as natural sympathies and antipathies, he opposed the adepts at a crucial point. The knowledge of the magi was the property of the initiated; it had to be acquired by direct teaching or by inspiration and passed on to the enlightened in riddles opaque to fools.[112] Agrippa says that he writes so as to "confound the ignorant," and Dee, despite his essays in general education, insisted on restricting and obscuring his higher knowledge. The printing press put these good men in a quandary. Should they use it to reach the enlightened everywhere, or should they continue the practice of the ancient Pythagoreans, passing on their wisdom only to initiates chosen by them? Dee characteristically shifts the problem to someone else: he gives his manuscript to the printer and beseeches him to keep the resultant book—in this case the incomparable *Monas*—from falling into the hands of the vulgar. His reasons are curious. Some people, unable to extricate themselves from the "labyrinth" of his thought, will "torture their minds in incredible ways [and] neglect their everyday affairs"; while others, "imposters and

111. Paolo Rossi, *Francis Bacon* (1968), pp. 14–33.
112. *Ibid.*, p. 30; Rossi (in Bonelli and Shea, ed., *Reason, Experiment, and Mysticism* (1975), pp. 256–264) has recently reemphasized the distance between Bacon and the scientific revolutionaries on the one hand, and the magicians on the other.

mere spectres of men," may deny the truths Dee revealed and question his integrity.[113]

Bacon reproved the hermetists for their secrecy, for their obscurity, and, above all, for their belief that knowledge advances by individual inspiration and limited circulation. Cardano had said that in making discoveries there is no need for partnership; Bacon insisted upon slow, plain, orderly, co-operative investigations. He would organize, set up committees, systematically extract the lore of philosophers, artisans, mechanics, and magicians. Unconfirmed effects, false claims, illusions, and fallacies would be exposed, rooted out, and thrown away with the refuse of the school philosophy.[114] Bacon's concept of an open, natural-magical research organization was an important ingredient of the Scientific Revolution. We find it already adumbrated in the little society della Porta set up in Naples; and we see it expressed in Kircher's museum, in the Accademia del Cimento, in Lana's Filoesotici, and in the royal societies of London and Paris.[115] There is no doubt that a denatured, democratic hermeticism and alchemy inspired some of the members of these groups.

Bacon and the natural magicians of the seventeenth century also differed from Yates's hermetists over the business of conjuring. Although the hermetist might draw back from invoking demons explicitly, he always flirted with spirits, for even the most benign hermetic rites implicitly addressed superior intelligences. Ficino, for example, strums Jovial music, burns Jovial incense, and contemplates Jovial pictures to prepare (he says) his own soul or an appropriately marked seal for the absorption of the beneficent rays of Jupiter; his idea is to bring his soul or the talisman into harmony with the planetary spirit so that they all might vibrate together. But as appears from the writings of Ficino's disciple, Francesco da

113. Dee, *Monas*, ed. Josten, *Ambix* 12 (1964): 151; cf. Nauert, *Agrippa*, pp. 17–18, 44.
114. Cf. Rossi, *Bacon*, pp. 32–34.
115. J. L. Heilbron, *A History of Electricity* (forthcoming).

Diacetto, the music, the incense, and the talismanic symbols were also part of a rite aimed to conciliate the planetary soul;[116] indeed, it could not be otherwise, for as St. Thomas had proved two hundred years earlier, symbols, songs, and prayers can secure the ends intended by the magician only if directed at superhuman intelligences. The use of intelligible signs to attract celestial powers presupposes an appeal to demons.[117] The demonic side of Ficinian astral magic is emphasized by the chief of the sixteenth-century magicians, Heinrich Cornelius Agrippa von Nettesheim, who openly describes, though he does not endorse, the standard methods for coaxing planetary spirits.[118]

One could give sound reasons for believing that astral intelligences could be summoned and commanded. First, analogy to the practices of the Church, to prayer, to the efficacy of the sacraments, and, above all, to the mystery of the eucharist: on every side one saw, or believed one saw, the intervention of superpowers mobilized by human operations. Agrippa made much of parallels in the workings of priests and magicians.[119] Second, there was excellent authority for the efficacy of hermetic astral magic, no less an authority than God; for the disciples of Hermes believed that their master had his wisdom from Moses, who had had his by divine revelation.[120] As Agrippa tells it, God presented Moses with

116. D. P. Walker, *Spiritual and Demonic Magic* (1958), pp. 14–17, 33, 198–199; Yates, *Giordano Bruno* (1964), pp. 62–82.

117. St. Thomas, *Summa contra gentiles*, III. c.103–105. Lefèvre made the same point, and Ficino himself had doubts about the legitimacy of talismans; yet Ficino, by endorsing the old Neoplatonists, especially Plotinus, tacitly advocated demonic magic. Walker, *Magic*, pp. 41–53, 167–170; but cf. Yates, *Bruno*, p. 91.

118. Walker, *Magic*, pp. 90–96; Nauert, *Agrippa*, pp. 139, 246, 255; Agrippa, *De occulta philosophia*, lib. i.c.71 and lib. iii.c.4.

119. Walker, *Magic*, pp. 36, 93–95; Pico made the connection even stronger by invoking magic to help prove the divinity of Christ (Yates, *Bruno*, p. 105). The Reformers, by rejecting much traditional Church magic, made the lot of extraclerical magicians harder. Cf. K. Thomas, *Religion*, pp. 303–305, 318–320, 326–327.

120. Yates, *Bruno*, pp. 1–19.

two revelations, one to be made public in Scripture, the other, too heady for the vulgar, to be passed down through a line of sages.[121] Why Moses chose to confide in an Egyptian on the eve of the flight of Israel is a hermetic mystery. That the revelation of Hermes does not derive from remote antiquity but from inspired Alexandrian Greeks given to pseudepigraphy was a discovery of the seventeenth century.

Among the hermetic books is one, the *Asclepius*, that teaches how to prepare objects for the acceptance of celestial influences and how to animate a statue with, or with the help of, astral intelligences.[122] The hermetist consequently accepted the existence of astral influences as part of his magical revelation. But he was not necessarily an enthusiastic astrologer. Agrippa, for example, always had doubts about traditional astrology, which he finally attacked as baseless and superstitious.[123] Likewise Ficino's disciple Pico della Mirandola, who had no difficulty with astral magic, devoted an entire book to the confutation of astrology. The resolution of the apparent paradox in their position turns on a distinction important for understanding the significance of Dee's astrological aphorisms.

The traditional astrology, the star lore the Latin Middle Ages took from the Arabs, was deterministic: the stars govern everything here below, and cannot alter the future they announce. The Catholic Church had utterly condemned this fatalistic doctrine in late antiquity; but when it returned in the twelfth century, enriched with the learning and mathematical obscurities of the Arabs, theologians thought rather to Christianize than to fight it.[124] The compromise worked out by Thomas Aquinas allows the stars full rights over matter, including the human body and its appetites, but exempts the soul. The doctrine turned out to be more useful to theologians than to common men, because, alas!, we only

121. Nauert, *Agrippa*, p. 47.
122. Yates, *Bruno*, pp. 37, 41.
123. Nauert, *Agrippa*, pp. 94, 199, 211.
124. T. O. Wedel, *Medieval Attitude* (1920), pp. 15–41.

too often allow ourselves to be carried away by passions driven by our appetites. "And so it is," says St. Thomas, "that astrologers are able to foretell the truth in the majority of cases. . . . " Only those able to rule their passions can escape them; only the wise man can win the endless struggle with the stars.[125]

This compromise did not satisfy the hermetists. Their revelation promised that they could not only evade, but also exploit, stellar influences. They opposed theoretical constraints on their freedom of action; man is a maker, a magus, a miracle; he operates as a god. "And so, O Asclepius," Hermes tells his disciple, "man is a *magnum miraculum*, a being worthy of reverence and honor," half divine, a cousin or brother of demons. What might he not accomplish through revealed wisdom and family connections? Pico sounds the same note in his *Oration on the Dignity of Man*, and goes on to emphasize freedom of the will and action.[126] Now these rhapsodies, however sunny and optimistic they may appear, in fact menace the very dignity they declare; for man's reason, which might be urged to be the first mark of his excellence, can neither discover nor explain the magic that hermetists applauded as evidence of his, or rather their, high state. Belief in the efficacy of unintelligible operations on the basis of revelation does not stem from confidence in the power of the human mind. And with the erosion of confidence in reason comes a willingness to accept reports of strange isolated occurrences, of singular events, of prodigies; one loses the basis of judgment, or, to use a favorite sixteenth-century image, one goes adrift without the only compass that can provide safe passage through seas of uncertainty and confusion. The connection between faith in Hermes' revelation, depreciation of reason, and gullibility appears clearly in

125. *Ibid.*, p. 68; T. Litt, *Les corps célestes* (1963), pp. 149–165, 201–213; cf. K. Thomas, *Religion*, pp. 428–430.

126. Yates, *Bruno*, p. 111; Pico's *Oration* is translated in Cassirer *et al.* ed., *Renaissance Philosophy* (1948), pp. 223–254.

Agrippa, who swings between occult philosophy and extreme scepticism, and finishes in fideism.[127]

We are left with the supposition that hermetic astral magic may be antithetic to the traditional rationalistic, computational, deterministic astrology. Astrologers may be closer to physicists than to magicians. The course of Dee's work confirms the conjecture. He began optimistic in his power to understand and develop a full mathematical astrology, free, as will appear, from appeal to demons. He presently lost his compass (which perhaps was never tightly secured) and wandered through Europe without reason or direction, asking recalcitrant angels for revelations that never came.

Yates's thesis sharpened

Yates is by no means the first modern historian to emphasize the importance of mystical Platonizing mathematics in the transition from peripatetic to early modern science. The thesis, as started by E. A. Burtt in 1924, emphasizes the importance of considerations of mathematical harmony and simplicity to Copernicus, Kepler, Galileo, Descartes, and Newton, considerations that might override objections to their novelties drawn from traditional physics, direct experience, or common sense.[128] The appeal to "reason" over "sense," to ideas over experience, is supposed to mark the difference between Plato and Aristotle. Galileo made it the ground of his spokesman Salviati's admiration for the Copernicans: "they have through sheer force of intellect done such violence to their own senses as to prefer what reason told them over that which sensible experience told them to the contrary."[129] The new criterion of truth is mathematical appropriateness: God is a geometrician, and so,

127. Nauert, *Agrippa*, pp. 200–203, 208–209, 218, 227.
128. Burtt, *Metaphysical Foundations*, 2nd ed. (1932).
129. Galileo, *Dialogue*, tr. Drake (1953), p. 328.

in our weaker way, are we. If we can but shed the prejudices of our senses we may recapture at least part of the Divine Plan, the figures, forms, and numbers—in short the mathematics—of creation. This is, as we know, a constant element in Dee's thought, and the inspiration, if not for his best mathematics, for his most ecstatic prose. "O comfortable allurement, O ravishing perswasion, to deale with a Science, whose subject, is so Ancient, so pure, so excellent, so surmounting all creatures, so used of the Almighty and incomprehensible wisdome of the Creator, in the distinct creation of all creatures: in all their distincte partes, properties, natures and vertues, by order, and most absolute number, brought, from *Nothing*, to the *Formalities* of their being and state."[130]

Burtt counts the early Copernican revolutionaries, Copernicus himself and Kepler, as Platonists chiefly because they took mathematical elegance as the touchstone of true physical theory. They may be admitted Platonists—or rather Neoplatonists—on another ground as well. Ficino had emphasized the prerogatives of the sun in words that virtually declared a heliocentric universe.[131] Copernicus recalled in *De revolutionibus* that the thrice-great Hermes had named the sun "a visible God"; for himself, he preferred to imagine it "resting on a kingly throne, govern[ing] the family of stars that wheel around."[132] These remarks, together with the emphasis on mathematics and the fact that Copernicus cultivated humanism during ten years' study in Italy, show that at some level—whether of enabling inspiration, congenial confirmation, or incidental decoration—Platonism figured in the composition of *De revolutionibus*.

130. "Preface," sig. *j^r.
131. P. O. Kristeller, *Philosophy of Ficino*, tr. Conant (1943): "Rest is more perfect than movement" (pp. 89, 173); the sun is perfect, first of its kind, the image of God (pp. 95, 98, 147). Can one resist the inference that it is at rest?
132. Copernicus, *De rev.* (1543), lib. i.c.10.

In the case of Kepler, the matter is plainer. He began his astronomical work with the conviction that God had the Platonic solids in mind when arranging the planets, and he came to his greatest discovery, the elliptical orbits, following an odd celestial dynamics derived from his belief that the sun, an image of God the Father, literally drove the planets around. In middle age he returned to the riddle of the number and spacing of the planets. To answer it, he took the notion of mathematical harmony to its literal extreme: he worked out that the Pythagorean music of the spheres, a music for the intellect, not the ears, is written with chords fixed by the ratios of the distances of the planets from the sun. The precise part assigned to each member of the choir depends upon its velocity. Kepler's search for harmonies caused him to examine many possible relations between the solar distances of a planet and its speed; and among the discoveries of his Pythagorean quest was the true and important "law" connecting the average solar distance (a) and the heliocentric period (T) of the planets ($a^3/T^2 = $ const.).[133]

Galileo had little patience with Kepler's procedures. He can therefore plausibly be held to be a Platonist only on the ground of his insistence that to understand the universe one must be able to read mathematics. For mathematics is the language in which the book of the universe is written; without a knowledge of its alphabet, of "triangles, circles and other geometrical figures," one cannot understand a word of it.[134] Many similar Platonic commonplaces may be gathered from Galileo's works. Historians of science have taken them very seriously. Under the influence of the late Alexandre Koyré it became fashionable to reckon Galileo's Platonism so strong as to have kept him from the experiments that had been considered his greatest glory.[135] But recently Stillman Drake

133. M. Caspar, *Kepler*, tr. Hellman (1959), pp. 60–70, 123–141, 264–289.

134. S. Drake, *Discoveries* (1957), pp. 237–238.

135. E.g., A. Koyré, *J. Hist. Ideas* 4 (1943): 400–428.

and others have restored him to his weights and inclined planes.[136] They may thereby save Galileo from Koyré's brand of Platonism, but—and this is a capital point—not necessarily from Yates's. Galileo's experimentalism, which embarrassed Koyré, is to Yates a natural development from the combination of elements first united in men like Dee. The novelty in her position, perhaps the chief cause of current interest in Dee, is the claim that hermetists helped to arrange the wedding of scholar and manipulator, of geometry and experiment.

We may consider Yates's claim as part of an explanation of the remarkable growth of interest in, and knowledge of, mathematics in the sixteenth century. There were of course other causes of this phenomenon than the program of the hermetists. One frequently invoked is the activity of the humanist translators, the new masters of Greek, who, having exhausted the ancient writings closest to their hearts, turned to making good Latin editions of the classic mathematicians. These translations, disseminated by the new printing presses, made accessible almost the entire range of extant Greek mathematics. Euclid, already available in incunabular printings of a medieval translation from an Arabic original, was reedited directly from the Greek into Latin and vernacular languages. Ptolemy's *Almagest* first came into Latin directly from Greek in 1528; Archimedes's *Opera* did the same in 1544, and the first four books of Apollonius's *Conics* in 1566.

Another cause for sixteenth-century interest in mathematics was utility. The new warfare based on gunpowder required the skill of practical geometers for both fortification and sieges, for constructing cannon-proof ramparts, running trenches, designing siege equipment. Dee's student, Thomas Digges, to name but one among many mathematicians, made

136. E.g., Drake, *Sci. Amer.* 228:5 (1973): 84–92, and *Isis* 64 (1973): 291–305; T. B. Settle in *Galileo*, ed. McMullin (1967), pp. 315–337. Cf. T. P. McTighe, *ibid.*, pp. 365–387.

part of his career as a military engineer.[137] Then there was civil engineering, the new architecture of Brunelleschi and Alberti, or rather the old architecture of Vitruvius. And, most important of all, the voyages of discovery and exploration, the lengthening trade routes to Africa and Russia, the governance and defense of vast empires made necessary the services of crowds of navigators, surveyors, and cartographers. It was this need that inspired Gemma's teaching and brought Dee to Louvain.

We have observed that the connection of mathematics with practice helped to earn it the disfavor of dons unwilling or unable to study it. The academic mathematician often met this prejudice by calling up the Greeks. Editors of Euclid, for example, liked to dwell upon the prerequisite for admission to Plato's Academy, a knowledge of geometry, without which one cannot fully grasp higher philosophical studies. Others met the prejudice head on, by extolling the incomparable Archimedes, the exemplar of the mathematician pure and applied. The sixteenth century took much from the ancient engineer of Syracuse. His elegant mathematics helped inspire its algebra; his studies of mechanics and hydrostatics gave it a model for quantitative descriptions of physical phenomena; and, above all, his traditional exploits, his destruction of the Roman fleet by burning mirrors, his single-handed moving of a great ship by a system of pulleys, taught it what a little mathematics, appropriately applied, could do.

Plato and Archimedes could, of course, be more than convenient symbols in academic disputes about the value of mathematics. They might also inspire fruitful lines of study. If Plato was Kepler's cicerone, Archimedes was Galileo's; while Kepler pondered the cosmological significance of the five regular solids, Galileo, who had learned his mathematics at the Florentine Accademia del Disegno along with painters, architects, and engineers, was composing essays on Archimedean

137. H. Webb, *Elizabethan Military Science* (1965), pp. 17–27, 64–65.

themes such as centers of gravity and floating bodies.[138]
Now Yates promotes Hermes as a third major inspirer and
director of mathematical studies in the sixteenth century,
particularly of studies that realized, or at least anticipated, an
amalgamation of mathematics and experiment. Among those
she likes to point to is Dee.

Students of the scientific revolution are therefore
challenged to discover how far (if at all) applied mathematics
and Yates's hermeticism interacted fruitfully in the sixteenth
century. Did a significant number of applied mathematicians
then derive their interest, direction, and legitimization from
the Pythagorean elements in hermeticism? What in their
writings is conventional or purely rhetorical? What part did
astrology and other quantitative pseudo-sciences play? Forty
years ago, E. W. Strong attacked some of these questions in
an attempt to answer Burtt's claim about the Platonic underpinnings of early-modern science.[139] Strong pointed out that
the chief contributors to pure mathematics in the sixteenth
century did not concern themselves much, if at all, with
numerology, while those most Pythagorean in spirit produced
relatively little. Hermes did not figure at all. Strong's polem-
ical purpose, however, took him beyond the mark; and he
ends by denying that Neoplatonic impulses assisted in any
way in the birth of modern science.[140]

The example of Kepler, or of the algebraist and cabalist
Stifel,[141] show that Strong went too far. Yates, too, has over-
played her hand. The secretive "experimentation" of her
magi, their tendency to accept reports of prodigies, and their
anti-rationalism ran against the currents of the scientific
revolution. Yet, the hermetists' emphasis on wisdom, on
studying subjects depreciated or ignored in the schools, and

138. L. Geymonat, *Galileo*, tr. Drake (1965), pp. 7, 19, 26.
139. Strong, *Procedures and Metaphysics* (1936).
140. E. Cassirer in *Galileo*, ed. McMullin, pp. 338–351.
141. *DSB*, XIII: 58–62; J. E. Hofmann, *Sudhoffs Archiv*, suppl. no. 9
(1968).

their promise of ultimate knowledge, may have provided psychological support for many to master and to promote neglected studies that had no necessary connection with magic. In this connection, the *Propaedeumata aphoristica* take on great interest. On the one hand, they represent an inclusive, quantitative, physical science, assisted by experiment and aimed at both understanding and control of natural processes. On the other, they echo the harmonies of the world and the sympathies among all things, and dimly announce the hermeticism that was to inform the *Monas*. The sources of Dee's aphorisms and the manner in which he reworked them consequently claim our attention.

⁓ II. *Propaedeumata Aphoristica*

I. ASTROLOGY AS APPLIED MATHEMATICS

The cosmology underlying Dee's astrological aphorisms may appear to be traditional. The framework of the whole is the Aristotelian earth-centered universe, finite in size and divided into two qualitatively distinct regions, the unalterable celestial and the corruptible sublunary. The devices of Ptolemaic astronomy regulate the motions of the heavenly bodies; the old doctrine of the four elements, with an admixture of alchemy, applies to the world beneath the moon. Neoplatonic sympathy ties everything together; according to Dee, it originates in the heavenly bodies and propagates through space as prescribed by the respectable medieval theory of the multiplication of species. When received on earth, the species or radiated sympathy gives rise to effects that depend upon its own nature and on that of its terrestrial absorber. Knowledge of elective sympathies, of the capacities of sublunary substances for soaking up or concentrating celestial species, enables the astrologer to set up as a magus or physician, and, perhaps, to alter the course of events.

Dee reworked these received qualitative propositions in the manner of the applied mathematician. He took as his guide the exemplar of all radiations, the pattern of multiplication of species, "the first of God's Creatures,"[1] namely

1. "Preface," sig. b.jʳ.

visible light. That immediately put at his disposal all geometrical optics then known. He came to assimilate invisible
species so closely with light that, in the "Preface," he defined
the art of perspective as the description of "all Actions, and
passions, by Emanation of beames perfourmed."[2] Not only
did optical theory give Dee a way to estimate the relative
strengths of astrological radiations in different planetary
configurations; it also enabled him to prescribe how to
confirm and extend the reach of the calculations by experiment. The main apparatus required was a parabolic speculum,
a burning mirror, supposed to concentrate all stellar emanations precisely as it intensified light.

Dee's approach to astrology, although not entirely
original with him, deviated in several ways from that of the
ordinary literature. There is no trace of the hermetic planetary souls, for example: stars are not intelligences to be
cajoled but unalterable sources of exploitable power; they
are not like people but like radiators; their influences may be
concentrated by optical instruments, not by songs, prayers
and incense. Again, Dee takes literally the demands of optical
theory; and since the intensity of radiation diminishes with
distance and increases with the size of the luminous source,
he insists that an exact astronomy, one that yields precise
values of the sizes and distances of all planetary configurations,
is a prerequisite to a competent astrology. The general
practitioner did not care for such refinements, which he could
neither use nor understand. We may take the grasp of
arithmetic of Andrew Borde, M.D., as representative: "Euery
one of the sygnes [of the zodiac] is deuydyd in lx degres," he
says, and "euery degre doth contayn .vi. mynutes [!];" from
which he works out that "yᵉ. zodiack [has] .CCC.vi. degres."[3]

A more instructive contrast with Dee is furnished by
Cardano, who published some 1173 astrological aphorisms in

2. *Ibid.*
3. Borde, *Pryncyples* (c. 1542), sig. c. 1.

1547. Cardano dismisses Trithemius and Agrippa as crackpots; like Dee, he has no place in his astrology for demons, an ineffectual race in his experience (his father had kept one) who have no effect on stars.[4] But Cardano, although a good mathematician and a professor of mathematics, had no program for a quantitative astrology. "Vita brevis, ars longa," he says, "experience is not subject to our will and judgment is difficult."[5] We must collect rules and guides from reading books and comparing nativities with biographies. To be sure, some knowledge of arithmetic and geometry is needed, but not much: the motions of the stars cannot be known perfectly, ephemerides need continual adjustment, and in any case no planet ever returns precisely to the same place in the heavens.[6] The surest guide is the study of genitures.[7] Cardano analyzes a great many of them, some of famous men; and he emerges with a welter of aphorisms impossible to reduce to general principles. "When the moon is under the sun she makes melancholics and great thinkers." "If the moon is in the second degree of Taurus, and either in quadrature or opposition to Jupiter and moving into trine with the sun, he [the subject of the geniture] will acquire not inconsiderable riches." "When the moon and Mercury are in conjunction in Taurus, he will be studious and even erudite."[8] Cardano alternately encourages and bullies the fledgling astrologer who may have difficulty swallowing several hundred of these morsels. "Constant repetition is necessary in this science," he says, for it is exceedingly hard. "[My] book is more difficult than Ptolemy's great composition [the *Almagest*] or the ars

4. Thorndike, V: 568–573; Calder, *Dee*, II: 275; Cardano, *Book of My Life*, p. 10. Cardano had a guardian angel, not a servile demon (*ibid.*, pp. 240–244).

5. *Aphor. astr.*, i.1 (this notation signifies "segmentum i, aphorism 1").

6. *Ibid.*, i. 7–8, 23.

7. *Ibid.*, i. 48, 50. Cardano also published 100 sample genitures, with analyses, in *Libelli quinque* (1547), ff. 102ᵛ–182ᵛ.

8. *Aphor. astr.*, ii. 30, 33; iii. 20.

magna [algebra], than Archimedes' spheroids, Aristotle's physics, or Plato's aenigmata."[9]

Dee's aphorisms differed from Cardano's in offering explanations rather than recipes and mathematics in place of mush. Among precedents for such an approach is Ptolemy himself, who tried to reduce astrological practice to intelligible rules based upon physical properties assigned to the planets. Envious or malevolent contemporaries accused Dee of plagiarizing from closer to home, from his friend Mercator or from medieval authorities, from old Urso of Salerno or from Alchabitius (al-Qabīsī, fl. 950), the author of an introduction to astrology often reprinted in the sixteenth century.[10] Had these peevish prattlers known that their man had borrowed a manuscript containing several works of al-Kindi (fl. 850), including *De radiis stellarum*, from Oxford in 1556, they could have made a better case.[11] Al-Kindi taught a straightforward determinist astrology, urged its merits as a mathematical science, and grounded it physically in the interplay of radiations continually and invariably pouring out of celestial bodies. He also recommended magic to accumulate useful rays, as had the Greek Neoplatonists from whom he took his theoretical underpinnings. Al-Kindi's views had enjoyed the advertisement of formal condemnation at Paris and Oxford during the thirteenth and fourteenth centuries, and of special attention in the widely read *Errores philosophorum* of Giles of Rome.[12]

Another astrological writer who doubtless poured his

9. *Ibid.*, ii. 126; iv. 83, 84.

10. *AT*, p. 56; Calder, *Dee*, I: 523–526, II: 255–258; *DSB*, XI: 226; Urso Salernitanus, *Aphorismen*, ed. Creutz (1936). Alchabitius also wrote a lost book on planetary distances; P. Duhem, *Système du monde*, II, 2nd ed. (1965), p. 53.

11. Corpus Christi MS 191, f. 89ᵛ (Bodleian); cf. Calder, *Dee*, II: 251.

12. Thorndike, I: 642–647. Giles of Rome, *Errores philosophorum* (c. 1270), ed. Koch and Riedl (1944), pp. xlvi–xlvii, liv–lvii, 47–57; M. T. d'Alvernay and F. Hudry, *Arch. hist. doct. litt. moyen âge* 41 (1974): 141–142, 149, 155–167.

influence on Dee was one Joannes Franciscus Offusius. Like Dee, Offusius understood that astrology could not become a science until it could compute celestial influences precisely; and having made this charge his own, he spent many years searching the world (in vain as it happened) for a suitable scientific collaborator. He stayed with Dee in 1553, and perhaps propounded a few of the three hundred astrological aphorisms that his host then confided to a manuscript now unfortunately lost.[13] He no doubt learned much from Dee, and perhaps even more after leaving him; and in an ephemeris for 1557 he was able to publish the results of his new system of astrology. As one might expect, he concluded that astral influences propagate in rays that obey mathematical laws; in particular, the influences diminish in power with distance from their sources according to curious laws of the inverse square and cube, and the average distances between earth and the planets, among other parameters of interest, spring from the geometry of the Platonic solids.[14] Offusius's system was more Pythagorean than Dee's, but otherwise much like it; Dee could not miss the affinity and cried plagiarism when, in 1570, a full account of his old collaborator's ideas was printed.[15] These priority disputes have at least one merit: by showing that Dee's program to base astrology on mathematical physics was not unique, they give his *Propaedeumata* the interest of a representative work.

Whoever tried to father Dee's aphorisms on Mercator came closest to the mark. Indeed, Dee acknowledged his obligations to his friends at Louvain in the dedicatory letter to the *Propaedeumata*, where he singled out not only Mercator and Gemma, but also Gaspar a Mirica (Caspar vander Heyden) and Antonius Gogava. Vander Heyden was a goldsmith and

13. *AT*, pp. 26, 75.
14. Thorndike, VI: 22–24, 108–111; M. J. Bowden, *Revolution in Astrology* (1974), pp. 78–89.
15. *AT*, p. 58; Offusius, *De divina astrorum facultate in larvatam astrologiam* (Paris, 1570).

engraver who had worked for many years for Gemma and who had probably taught Mercator his art; he made instruments too and may have instructed Dee in the manufacture and use of astronomical and optical apparatus.[16] Gogava was a twenty-year-old prodigy who had just finished a translation of Ptolemy's *Tetrabiblos* when Dee arrived in Louvain.[17] It is the first complete Latin version made directly from Greek. In the same volume, Gogava gave two medieval tracts, one on conic sections, the other on burning mirrors; he thereby provided the modern astrologer with all he needed to make use of vander Heyden's practical instruction, and to philosophize in the foreign, or at least the Louvain, manner. Dee says that the "earnest disputations" of Gogava did much to "provoke" him to a serious study of astrology.[18]

As for Gemma and Mercator, both firmly believed in the power and reach of astral influence. Gemma, who with his fellow physicians held astrology to be essential to correct medical practice, endorsed Gogava's translation of Ptolemy for its up-to-date practical utility.[19] Mercator, although a cartographer by profession, preferred philosophy, and understood the structure and operation of the universe. The sun and stars occupy the chief place in his world. They are perfect and unchanging, as Aristotle taught, the most noble instruments of God, sympathetically and genetically tied to the sublunary world. If they were to alter it would go hard with us, for it is their business to constitute, preserve, and perpetuate things here below, "ad maturandos inferioris mundi

16. Cf. A. de Smet, *Der Globusfreund* 13 (1964): 32–37, and in Soc. belge d'études géogr., *Bull.* 32 (1963): 34–35; R. Haardt, *Imago mundi* 9 (1952): 109–110.

17. Gogava, letter dedicatory, in Ptolemy, *Opus quadripartitum* (1548); J. Verheiden, *Vita* (1596), pp. 2–3.

18. *AT*, p. 5; "Preface," sig. b.iiij^r. Dee made astronomical/astrological observations at Louvain in August and November, 1548 (Calder, *Dee*, II: 107).

19. F. van Ortroy, "Bio-bibliographie de Gemma Frisius," (1920), p. 23. Gemma contributed a preface to Gogava's book.

foetus, eisque obstetricandum." Mercator's sunny conviction that stellar influences are inherently good (evil arises from irradiating corrupt or inappropriate matter) recurs in Dee.[20]

The sizes and distances of the planets, the main parameters of Dee's mathematical astrology, were also of vital interest to Gemma and Mercator. The hope of learning something new and precise about these parameters, about a world "hitherto described [only] within uncertain limits," had aroused in Gemma a great desire to read the work of Copernicus. He had gone so far as to urge Copernicus's bishop to shake the long-awaited book out of him. "I do not care [Gemma wrote] whether he says that the earth revolves or stands still . . . ; the only bad thing is delay."[21] As will appear, planetary distances are determined more directly from Copernicus's theory than from Ptolemy's. It was precisely on this head that Gemma later argued the superiority of Copernicus's hypotheses: "they do not bring anything absurd into natural motions, since they allow us a fuller knowledge of the planetary distances than do [Ptolemy's]."[22]

Gemma's endorsement of the new astronomy served as preface to a set of ephemerides calculated by Joannes Stadius after the *Prutenic Tables* of Erasmus Reinhold, who had worked from the parameters and hypotheses of Copernicus. Stadius's work appeared in 1556; a year later, Dee's friend John Feild brought out similar tables, with a similar preface by Dee. In the same vein as Gemma, Dee emphasized the inaccuracies and stupidities of the older ephemerides and the promise of

20. A. de Smet in *L'Univers à la Renaissance* (1970), pp. 24–28, and *Duisb. Forsch.* 6 (1962): 35–36; Mercator, letters of 1581 and 1585, in *Corresp.*, ed. van Durme (1959), pp. 165–166, 192–193; Mercator, *Atlas* (1595), pp. 9, 11, 17–19. Mercator's celestial optimism may be found in many earlier writers, e.g., Plotinus (Calder, *Dee*, II: 248).

21. Gemma to Joannes Dantiscus, July 1541 and Aug. 1543, in van Ortroy, "Bio-bibl.," pp. 409–410, 413.

22. Gemma to Johann Stadius, 28 Feb. 1555, in Stadius, *Ephemerides novae* (1556); cf. G. McColley, *Isis* 26 (1937): 322–325.

the new, computed according to the "divine studies of Copernicus."[23] As for Copernicus's hypotheses, Dee declined to discuss them; he may have considered them to be no more than the best available calculating device, as did Reinhold and perhaps also Feild.[24] But we may be sure that Dee shared Gemma's interest in the computation of planetary distances according to the new astronomy. In the "Preface" he gives Copernicus's values for the solar and lunar distances.[25] Moreover, he there makes the first service of astronomy the "certification" of "the distance of the Starry Skye, and of eche *Planete* from the Centre of the Earth; and of the greatness of any Fixed Starre sene, or *Planete*, in respect of the Earthes greatness." A traditional definition would rather emphasize the planetary motions.

And why the consuming interest in dimensions? Because they give us the thickness of the "heauenly Palace" in which the planets do their exercises, and "meruailously perfourme the Commandement and Charge given to them by the omnipotent Maiestie of the king of kings."[26] As we know, Dee contributed to the technique of determining planetary distances, which he advertised as "a most beautiful part of philosophy, and most necessary to man."[27] His method is

23. Dee in Feild, *Ephemeris anni 1557* (1556). Cf. *Propaedeumata*, Aph. XCVI, which appears to endorse a geocentric version of Copernicus's notion that the eccentricity of the earth's orbit changes slowly in time.

24. Cf. Johnson, *Astronomical Thought*, pp. 134–135; A. Birkenmajer in *La Science au seizième siècle* (1960), pp. 169–180; R. S. Westman in *Copernican Achievement*, ed. Westman (1975), pp. 285–345. Recorde also mentions Copernicus favorably, though without endorsement, in 1556, in *Castle of Knowledge*. L. D. Patterson, *Isis* 42 (1951): 218; Kaplan, *Recorde*, pp. 153–159. An even-handed denial of the truth of every mathematical astronomy occurs in the *Planetologia* (1551), p. 14, of Dee's acquaintance Mizauld: "Qui omnes [epicycles, eccentrics, etc.] imaginarij prorsum habendi sunt, sed tamen artificiosè ad motus demonstrandos, & eliciendos, ut dictum est, excogitati."

25. "Preface,"sig. b.ij^r; chap. II:3, below; M. Bowden, *Revolution in Astrology*, p. 69.

26. "Preface," sig. b.ij^r.

27. Dee, *Parall. comm.*, sig. D.iiij^v (recte D.iij^v).

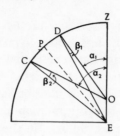

FIG. 6. Dee's procedure for determining parallax of a circumpolar object.

ingenious if seldom practicable. It requires measurement of the zenith distances a_1 and a_2 ($\angle$ZOD and $\angle$ZOC in Figure 6), and the calculable polar distance ϕ. Dee proposes to find the standard measures of distance, the parallaxes β_1 and β_2 ($\angle$EDO and $\angle$ECO), by a geometrical solution, borrowed from Regiomontanus, of what amount to the simultaneous equations $\sin a_1 : \sin a_2 = \sin \beta_1 : \sin \beta_2$ and $\beta_2 - \beta_1 = a_2 - a_1 - 2\phi$.

Dee could have taken a second principal ingredient for his astrology from his friends at Louvain. The tract on burning mirrors published by Gogava was often wrongly attributed to Roger Bacon; and it may have been the concern of Gemma's group with Bacon that brought Dee to recognize the bearing of the work of his medieval countryman on the computation of astral influence. By 1556, medieval optics was the subject best represented in Dee's collection of manuscripts, which included several tracts by Bacon, Alhazen, and John Peckham, the author of the standard *Perspectiva communis*.[28] Dee's interest in Bacon was to extend well beyond optical theory. He later developed and, as we have seen, algebrized, Bacon's idiosyncratic calculus of the graduation of qualities, and he undertook to clear him of charges of practicing unwholesome magic and devilish arts.[29] His aphorisms in particular

28. James, *List*, pp. 11–14.
29. Chap. I:2 above; Dee, "Speculum unitatis, sive apologia pro fratre Rogero Bachone" (MS of 1557), *AT*, p. 26. Cf. Yates, *Rosicrucian Enlightenment* (1972), p. 107, and F. Alessio, *Mito e scienza* (1957), pp. 13–14.

owe much to Bacon's notions of the propagation of force.[30]

The occasion for the aphorisms, in which Dee "mathematically furnished up the whole Method of astrology,"[31] was an inquiry from Mercator about the progress of his studies. What better answer than a little book to commemorate the lively discussions of astrological matters from which he had profited at Louvain? That, at least, is the motivation Dee gives for his first essay into print. But we may suppose that he also fired off the aphorisms to anticipate Offusius. And he might have wished to do his part in silencing those loud local critics then recently blasted by Leonard Digges, those "busy byghtinge [backbiting] bodyes" who rejected "the secret truthes" of astrology. Digges had discovered that these bodies byghted out of "ignorantie, the grete enemie of all pure learning."[32] John Feild took the same tack: the ignorance and bumbling of the vulgar chart-makers and nativity-casters have brought disrepute and even ruin to the discipline.[33] Dee's aphorisms offer a stronger defense. Although full of the usual sneers at the ignorant and uninitiated, they set out a rational astrological method open to anyone prepared to study them.[34]

30. Burton, *Anatomy*, ed. Dell and Jordan-Smith (1927), p. 422; Clulee, *Glas*, pp. 104–114, 123–124.

31. "Preface," sig. b.iij^r.

32. Leonard Digges, *A Prognosticon* (1555), p. xv. For Elizabethan opinion about astrology, see P. H. Kocher, *Science and Religion* (1953), chap. 10, and K. Thomas, *Religion*, chaps. 10–12. In general, the most prominent defenders of the art were the leading "scientists," that is, physicians and mathematicians. Kocher, *Science and Religion*, pp. 202–203; Kaplan, *Recorde*, p. 120.

33. Feild, *Ephemeris*, Preface; cf. Thorndike, VI: 99–144, and Cardano, *Aph. astron.*, iii.105: "Multi reliquunt Astrologiam, ut vanam, alij ut uncertam nobis. Qui autem haec quae hic tradita sunt non intelligent, ut difficillima destituent."

34. Dee was later driven to dismiss criticism of his *Monas* with the old saw, "scientia non habet inimicum, nisi ignorantem" (*AT*, p. 55): "Universitie-Graduates of high degree, and other gentlemen ... therefore dispraised it, because they understood it not" (*ibid.*, p. 10). Cf. Dee's disposal of Libavius, *ibid.*, p. 77.

It appears that Dee composed his aphorisms, or rather threw them together, in his customary hurry, perhaps quarrying them here and there from the three hundred that he and Offusius had discussed in 1553. Consequently, the whole wants system. The following pages present a systematic commentary on their physics, astronomy, and astrology, together with an outline of the necessary technical foundations. Nothing exceeds the grasp of a sixteenth-century undergraduate, even of Oxbridge. References to the aphorisms are given in Roman numerals.

2. PHYSICS IN THE *PROPAEDEUMATA*

The universe sketched in Dee's aphorisms centers upon the traditional stationary earth, whose "privilege" it is to be bathed in celestial radiations as the stars and planets wheel around (XCI). There is no doubt that, despite his interest in Copernicus, Dee took the apparent motions of the stars to be the "true ones": not only does he say so *expressis verbis* (LXXXIV), but otherwise his frequent identifications of heavenly motions as indicators and instruments of the universal harmony make no sense (LXXV, LXXVIII, LXXXVIII).

The earth, together with its oceans, its airy atmosphere, and, beyond that, its envelope of fire or ether, constitutes the elemental region, where everything is a product, example, or reproduction of the celestial harmonies (CXIV). This dominion includes man himself; for has Hermes himself not written, "nothing happens to man without cosmic sympathy" (CXIX)? The envelope of fire stretches to about fifty-two terrestrial radii (t.r.), which Dee, following Copernicus's estimate, made the distance of the lunar perigee.[35] The moon herself is preeminently the principle of moistness (CIII–CV).[36]

---

35. Dee, "Preface," b.ij^r; cf. J. Henderson in Westman, ed., *Copernican Achievement*, p. 118.

36. Ptolemy, *Tetrabiblos*, ed. Robbins (1940), p. 34.

Beyond the moon lie the orbits of Mercury, Venus, and the sun, in that order. The sun is, of course, the principle of light and heat (XCV, CII). Since, as Aristotle taught, heat and moisture are the efficient causes of growth, the sun and the moon are (after God) "the chief and truly physical causes of the procreation and preservation of all things" (CVI). As for the planets beneath the sun (Mercury and Venus) and those above him (Mars, Jupiter, and Saturn), each has a characteristic virtue (L, XC), which Dee does not bother to specify. Beyond Saturn, at a distance of about 20,000 t.r. (some 80 million miles), stands the sphere of the fixed stars or eighth heaven. Each star has its peculiar power, which Dee also allows his readers to discover for themselves.

Multiplication of species

Although he runs lightly over the natures of the celestial influences, Dee takes great trouble to describe their mode of propagation. As a traditional physicist he refuses to admit the possibility of action at a distance between material bodies: where such an action appears to obtain, as in the drawing of iron by a lodestone, we must assume the existence of an unseen link. "In the magnet God has offered to the eyes of mortals for observation qualities which in other objects he has left for discovery to the subtler research of the mind" (XXIV).[37] We easily infer from the operation of the lodestone, first, that magnets diffuse their power throughout the surrounding medium to a certain distance, their *sphaera activitatis*; second, that this power does not work visibly upon the

37. Dee's annotated copy of the first printing of the great medieval treatise on magnetism, Petrus Peregrinus's *De magnete*, ed. A. P. Gasserus (Augsburg, 1558), is in the British Library. Dee has underlined Gasserus's blast at scholastic occult qualities ("parum firmum Scholasticorum perfugium") and noticed his endorsement of Copernicus (sigs. A.iii^v, B.iii^r); and he has doubly marked Peregrinus's guess that the entire lodestone imbibes "virtue" from the entire heavens, poles from poles, and the rest from the rest (sig. D.ii^v).

medium, but only upon very special sorts of mixed bodies, namely bits of iron and other lodestones; and, finally, that the outcome of the interaction, "local motion" (*una ad aliam localiter accurrit*, X) depends upon the nature of the recipient, iron coming to the magnet, lodestones approaching or fleeing one another according to the orientation of their poles. Note that the interaction of lodestones is reciprocal, each activating the other to local motion.[38]

The medieval technical term for the diffusion of power through a medium was "multiplication of species." As one sees from the magnetic case, a species is not a material emanation but a peculiar condition of the medium impressed upon it by an active source.[39] "Species," says Bacon, "is the first effect of any natural agent." He has the concept from the leading Arab authority on optics, Alhazen, perhaps via Grosseteste, who built his natural philosophy on the physics and metaphysics of light.[40] Optical theory was the guiding analogy in the design of the doctrine of multiplication of species. One generalized from the distinction between *lux*, or light in the incandescent body, the *qualitas corporis lucentis*, and *lumen*, the *species lucis*, light in the medium, which becomes visible to us in passing through such substances as stained glass.[41]

Another example may help. Nineteenth-century physicists approved a theory that placed the seat of electrodynamic action in a hypothetical medium—the "ether"—the condition

38. This theory, which comes from Averroes, recurs in Roger Bacon's and Thomas Aquinas's commentaries on the *Physica*: Bacon, *Opera hactenus inedita*, ed. Steele *et al.* (1909–1940), XIII: 338–339; Thomas Aquinas, *Commentary*, ed. Blackwell *et al.* (1963), p. 433. Cf. A. C. Crombie, *Robert Grosseteste*, 2nd ed. (1962), pp. 211–212; J. Daujat, *Origines* (1945), II.

39. *Opus maius*, ed. Bridges (1900), II: 502–504; cf. *ibid.*, I: 111 (*The Opus maius*, tr. Burke [1928], I: 130).

40. Crombie, *Grosseteste*, pp. 104–116, 139–162; D. C. Lindberg, *Isis* 58 (1967): 335.

41. Bacon, *Op. maius*, ed. Bridges, II: 409. Cf. S. Vogl in Little, ed., *Bacon Essays*, pp. 207–209, and *Die Physik Roger Bacos* (1906): 40–50.

of which depended upon the number and the motion of charged bodies contained in it. A magnetic field might be represented by a vortical rotation of the ether around magnetic lines of force.[42] This rotation is a mechanical representation of the multiplication of magnetic species. The example fails, however, in two important respects. First, as a mechanical model it strays from the spirit of medieval natural philosophy, which saw neither the point nor the possibility of mechanical reductionism. Second, it postulates a special mediating ether, a substance not perceptible by the senses, whereas the medieval philosophers ascribed the necessary properties to the things about them. The medium for the propagation of *lumen* is anything transparent, like air, water, or glass; magnetism will seize on any medium but garlic juice;[43] and celestial influences propagate to some degree through everything here below (XXV).[44]

Dee, like Bacon, endows all things actively or actually existing with the power to send forth species (IV). Both substance and accident enjoy the power, spiritual substances as well as physical, all in proportion to ontological rank and degree (V).[45] The different emanating powers express themselves in species of different strengths (VI).[46] Regardless of power, the manner of propagation is the same, namely "spherical": from each point of the agent rays pour forth in straight lines in all directions (IV). ("Ray" signifies "species fashioned into a straight line by extension," that is, neither an object nor a state, but the mathematical construct of

42. For examples see E. T. Whittaker, *History*, I, 2nd ed. (London, 1951), chaps. 8–9.

43. A. von Urbanitzky, *Elektricität und Magnetismus* (1887), p. 22; Thorndike, VI: 282, 317, 420.

44. Bacon, *Op. maius*, ed. Bridges, II: 478.

45. *Ibid.*, pp. 418–427.

46. The caveat (V) that the nobler the thing the less "complete" its species (i.e., the less powerful in respect to the rank of its source) is necessary to keep celestial bodies from transmuting the elements into substances of higher degree.

geometrical optics.[47]) The spherical form of propagation, being the most perfect, is appropriate to light (XV) and to other species, which can flow either with light or without it (XIV); but without light, other forms—Dee has in mind principally the astrological qualities of heavenly bodies—can do nothing (XXII). Hence the very great importance of knowing when, and for how long, a given planet will remain above any assigned horizon (LV, LVI).

The rays of species obey the usual optical laws.[48] In particular, stellar rays bounce off the primum mobile (the sphere driving the eighth heaven) as from a perfect concave mirror (XXVIII), and bend in the earth's atmosphere in the same manner as light; consequently stars and planets just rising or setting may by refraction, and those near the nadir may by reflection, exert an influence on human affairs (XXIX). Similarly nonvisible species, in the same way as light, act more strongly the closer source and receiver (XLIII), as well as the nearer the angle of incidence to the perpendicular (XXXII). It is therefore of the first importance to know the distances of the luminaries from the earth.

Assuming the distances known from astronomy (XL), Dee (and Bacon) parade some elementary geometry to suggest the effect of the relative sizes of earth and stars on the magnitude of stellar influence. In Figure 7, V is a terrestrial point (the "vertex"), CV the central ray, AVB the luminous cone, ACB the luminous base, and AB the bounding circle (XXXIII). Rays within the cone are stronger the closer they are to the central ray (XXXIIII). (In the case of light, the central ray, CV, so overpowers the others that it alone makes an impression on the eye; otherwise, according to the received optical theory, a confused image would result from the coincidence at the

47. Lindberg, *John Pecham* (1970), p. 109; Bacon, *Op. maius*, ed. Bridges, II: 458–459. Cf. Pecham in Lindberg, *Pecham*, p. 108: "Omne corpus naturale visibile seu non visibile radiose virtutem suam in alia porrigere."

48. *Op. maius*, ed. Bridges, II: 465–468, 481–486.

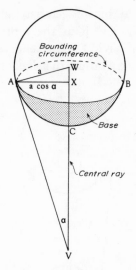

FIG. 7. Geometry of species rays from the illuminated surface ACB of a star of radius a and center W as seen from the "vertex" V.

same point V of the eye of rays from different points of the object.[49]) Dee takes as measure of the strength of the cone of rays the angle a, where $\sin a = a/r$, $a = $ AW being the star's radius and $r = $ VW being the distance of its center from the vertex (XXXVII). The radius of the base of the cone, AX $= a \cos a$, is less than a (XL), but grows with r; for $r \gg a$, AX $\simeq a(1 - a^2/2r^2)$. Hence Dee's claim that the base increases with the distance (XLI). He therefore faces two possibilities for the dependence of the strength of the radiation on r: on the one hand, as experience shows in the case of light and heat, strength is greater the smaller r; on the other hand, the base AX (and hence the size of the source) increases with r, and (another augmentation of strength) the rays within the cone come closer into parallel with the central one. Dee arbitrarily declares in favor of the first alternative (XXXVIII, XLIII), although he later hedges his bet (LXXXIX).

49. *Ibid.*, pp. 37–42, 511–516; Lindberg, *Pecham*, p. 38.

There is a final bit of geometrical mystification. In Figure 8, AC < BD, CD joins the centers and AB is the common tangent. If circle D represents the earth, less than half of it is illuminated by more than half of the star C (XXXV);

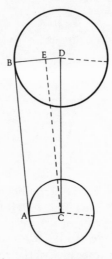

FIG. 8.   Ratio of illuminated to illuminating surfaces of two spherical bodies, one (BD) larger than the other (AC).

for ∠ACD is greater, and ∠BDC less than 90°, as appears from dropping the perpendicular from C on BD. (The point of intersection, E, must lie between B and D since BD > AC.) Conversely, C being the earth and D the star, less than half the star shines on more than half the earth (XXXVI). The phenomenon, and its dependence upon the separation of the bodies CD, demands the close attention of the astrologer: he who wants to follow Dee must determine, "with the greatest diligence" and for all possible cases, the ratio of the illuminated surface of the earth to the luminous surface of the illuminating star (XXXIX).[50]

50. Cf. *Op. maius*, II: 494–501.

Catoptrics

Catoptrics, which Dee recommends for its very great utility in searching out the "hidden virtues of things" (LII), is the study of the reflection of light. His knowledge of optics, which he, Recorde, and Leonard Digges had from Alhazen via Bacon and Pecham,[51] included a theory of the focusing of a parallel bundle of rays by spherical or parabolic surfaces. In the case of a concave spherical mirror, all rays parallel to a given diameter (QA in Figure 9) and equidistant from it pass,

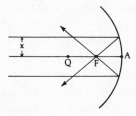

FIG. 9.   Reflection from a concave mirror of rays at a distance x from diameter QA; F is the focus.

after reflection, through the same point F. Rays at a greater distance than x come to a focus between F and A, and even beyond A.[52] Consequently, as Bacon observed, a spherical mirror provides only a "moderate convergence" of light and heat, albeit enough to ignite dry leaves.

A much grander fire can be made with a parabolic mirror, which reflects all rays of an incident bundle parallel to its axis toward the same axial point. Bacon knew only of the possibility of such mirrors, which he thought weapons fit

51. E. Wiedemann, *Ann. Phys.* 39 (1890): 128–129, and in Little, ed., *Bacon Essays*, pp. 185–203; Crombie, *Grosseteste*, p. 279.

52. Pseudo-Euclid in "Alkindi, Tideus und Pseudo-Euclid," ed. A. A. Björnbo and S. Vogl (1912), pp. 105, 115–117; Lindberg, *Pecham*, p. 209: "All rays reflected from such a [parabolic] mirror converge at one point so as to rarefy and ignite the air. In concave mirrors of spherical shape, on the other hand, reflection of all rays to one point occurs only from a single circle, and therefore concave spherical mirrors kindle fire weakly."

for Antichrist;[53] Dee understood them better, perhaps from Alhazen's explanation as published by Gogava, and he wrote a competent book about them—or rather a paraphrase of Gogava's tract—the same year that he composed our aphorisms.[54] He certainly had an eye to their power when he advised that the expert in catoptrics could "by art imprint the rays of any star much more strongly upon any matter subject to it than nature does herself" (LII).

The competent catoptrician is not limited to baking things at the foci of his mirrors. He can also cause a planet to stand off a very great distance from the earth, "and that," Dee puts in mysteriously, "within the space of a few days." Then, "in the winking of an eye, [he] may be able to draw it, as it were, to a new perigee" (LXXXIX). All this may mean only that with a few days' labor an adept can make himself a large concave mirror with which he can simulate the radiation that would be received were the planets further removed from the earth.[55]

If Dee did intend to augment and decrease the intensity of radiation catoptrically, he could have operated in something like the manner sketched in Figure 10, where BA is the mirror, Q its center, F its focus (whence AF = FQ), and P the radiating planet. Place a specimen or talisman XYZZ'Y'X' (XY = YZ) as shown, and shield YZ from the incident radiation by an appropriate absorber of celestial influence. Then, even ignoring absorption by XYY'X', less radiation will fall

53. *Op. maius*, ed. Bridges, I: 115–116 (Burke, I: 134–135). Cf. Vogl, *Die Physik*, pp. 67–72.

54. Wiedemann, *Ann. Phys.* 39 (1890): 110–130; *AT*, p. 26; N. H. Clulee, *The Glas of Creation* (1973), p. 125. A part of Dee's manuscript, "De speculis comburentibus libri 5, . . . circa illam coni recti atque rectanguli sectionem quae ab antiquis mathematicis parabola appellabatur," survives in Cotton Vitellius c. VII, ff. 275–308 (British Library); it scarcely advances beyond definitions.

55. Dee may have had this game from Bacon, who of course did not tell how to play it: Roger Bacon, *Letter*, tr. T. L. Davis (1923), p. 28.

upon Y'Z' than upon the equal surface XY. Move the talisman toward P, and its backside will intercept still fewer rays; and, in general, one can find a position at which the ratio of the irradiation of Y'Z' to that of XY has any assigned value less

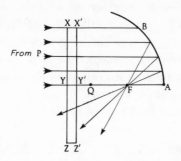

FIG. 10. The application of catoptrics to graduating celestial influences; Q is the center and F the focus of the spherical mirror BA; XYZZ'Y'X' is the talisman, P the radiating object.

than one. Compare Dee's boast that he can decrease the sun's heat (XCIX). No doubt decreasing the intensity of radiation by reflection is what Dee meant by causing a planet to stand off catoptrically. His purpose? To permit quantitative comparison between the radiations of the several planets.

The second part of the operation, drawing in the planet "in the winking of an eye," probably amounts to no more than moving the talisman to the focus: an area about Y' would then be irradiated much more intensely than any equal area of the surface XY. In a word, the reflector would act like a burning mirror, again to make possible accurate comparison of planetary influences. Once more consult XCIX, where Dee offers to augment the sun's heat to any degree within wide limits.

Play with large concave reflectors has a long and dark history. Everyone knows the fable of the mirror of Archimedes, which burnt the Roman fleet out of the harbor of Syracuse. More to our present purpose would be the great magnifier of

Alexandria, in which ships could be seen clear across the Mediterranean, and the speculum with which Julius Caesar, according to an old story retailed by Bacon,[56] spied upon the British from the coast of Gaul. Neither instrument could have worked, unless—it is the reservation of Gaspar Schott, S.J.— they performed by magic.[57] Still, like the fable of Archimedes, they say nothing impossible in principle, but exaggerate effects easy enough to produce *en petit*. It is theoretically possible, for example, to see distant objects enlarged with the aid of a single concave mirror.

In Figure 11, the distant object XY, say a church steeple

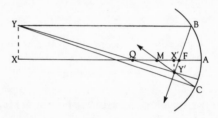

FIG. 11. Magnification of a distant object in a large concave mirror BAC, center Q, focus F; the eye is placed at M.

several miles away, throws an inverted real image X'Y' close to the focus F of the mirror BAC. The observer's eye rests at M between F and the mirror's center Q. The apparent size of the steeple is measured by the small visual angle it subtends at the eye, namely $\phi_{obj} = \angle XMY \simeq XY/MX$, or, because MQ is very much smaller than QY, $\phi_{obj} \simeq XY/QX$. The apparent size of the image, $\phi_{im} = \angle X'MY'$, is roughly $X'Y'/MX'$. Plainly $\phi_{im} > \phi_{obj}$, since $\angle X'MY' > \angle X'QY'$. The amount of magnification may easily be calculated. Let R be the radius QB. Owing to the great distance of the steeple in comparison

56. *Op. maius*, ed. Bridges, II: 165 (Burke, II: 581); Bacon, *Letter*, p. 29. Cf. Wiedemann, *Ann. Phys.* 39 (1890): 116; G. Schott, *Magia universalis*, 2nd ed. (1677), I: 413–420.

57. Th. H. Martin, *Bull. bibl. fis.* 4 (1871): 181–189; Anonymous, *Phil. Mag.* 19 (1804): 245–246; Schott, *Magia*, I: 443.

with R, X′ lies very close to F, or QX′ $\simeq$ R/2. But ϕ_{obj} = X′Y′/ QX′ = 2X′Y′/R, and ϕ_{im} = X′Y′/MX′; hence,

$$\phi_{im}/\phi_{obj} = R/2MX′.$$

Magnification occurs when R > 2MX′, and increases with R. If we suppose MX′ to be 10 inches, the shortest distance for clear vision for a normal eye, an effective catoptrical telescope must have a radius of many feet. Unfortunately, if it were to work as described the observer must have a transparent, nonrefracting head. The more usual opaque variety can also succeed if one inserts a small plane mirror, tilted so as to throw the image off the optical axis YA, in place of the eye between M and F. Della Porta described a similar arrangement to reflect the real image of a concave mirror to the eye. One might also succeed by placing the eye off the optical axis YA. A natural magician of the Enlightenment, Bonaventure Abat, did so using a concave mirror with a radius of curvature of 44 feet. And he pointed to the curious case of the canon of Erfurt who once saw a life-size image of a crucifix standing in mid-air. The canon had the sense to summon a theologian who was also a physicist, Joannes Zahn, who explained to him that he had seen the image of a small distant crucifix magnified by a "plane" mirror that had become distorted into a piece of the surface of a sphere of large curvature.[58]

It was perhaps with such mirrors that Dee expected military gentlemen to "make true report, or nere the truth of the numbers and Summes, of footmen or horsemen, in the Enemyes ordring."[59] He called this device a perspective glass, perspective, as we know, dealing with all radiations, and glass

58. Della Porta, *Magia naturalis* (1589), lib. xvii.c.5; Anon., *Phil. Mag.* 19 (1804): 181–182; Abat, *Amusemens* (1763), pp. 366–373; Zahn, *Oculus artificialis teledioptricus* (Wurzburg, 1685), quoted by Abat, *ibid.*, pp. 407–408; cf. D. Baxandall in Optical Society, *Transactions* 24 (1923): 305–307. One can also make a telescope of a single lens; see the report of the experiments of E. W. Tschirnhaus in Paris, Académie des sciences, *Histoire* (1700): 131–134.

59. Dee, "Preface," sig. a.iiij^v.

being a "generall name, in this Arte [Catoptrike], for any thing, from which, a Beame reboundeth."[60] These glasses had many imperfections both optical and tactical.[61] With uncharacteristic candor Dee almost allowed as much: although contemporary glasses were "wounderfully" helpful, he trusted to a "more skillfull and expert" posterity for instruments "to greater purposes, then in these dayes, can (almost) be credited to be possible."[62]

It may be that moderate success achieved with catoptric magnifiers inspired Recorde and Leonard Digges to try what might be done with other combinations of mirrors and even of mirrors and lenses. With some system of "proportionall glasses duely situate at convenient angles" Digges is said to have discerned what went on in private places seven miles off.[63] On the strength of this evidence, which we owe to Digges's son Thomas, Leonard Digges has been put forward as the inventor of the telescope, both refracting and reflecting.[64] Regrettably, neither of the Digges, nor Dee, all of whom were regarded as leading authorities in optics,[65] left us an instrument, or even an exact description of one.

The best we have from Dee is an account of fencing with a mirror belonging to his friend Pickering:

> If you, being (alone) nere a certaine glasse, and proffer, with a dagger or sword, to foyne at the glasse, you shall suddenly be moued to giue backe (in manner) by reason of an Image, appearing in the ayre, betwene you & the glasse, with like hand, sword or dagger, & with like quicknes, foyning at your very eye, likewise as you do at the Glasse.

The same riskless duel figures prominently in della Porta's bag of tricks, whence it entered the natural-magical repertoire

60. *Ibid.*, sig. b.j^r–b.j^v. Cf. C. de Waard, Jr., *De uitvinding* (1906), p. 72.

61. Cf. W. Bourne in *Rara mathematica*, ed. Halliwell-Phillipps, 2nd ed. (1841), p. 47; R. T. Gunther, *Early Science* (1967), II: 291–293.

62. "Preface," sig. b.j^r.

63. Thomas Digges, *A Geometrical Practice Named Pantometria* (1571), quoted in Gunther, *Early Science*, II: 290.

64. Gunther, *Early Science*, II: 293.

65. Bourne in *Rara mathematica*, p. 45.

of the seventeenth century. Della Porta's description is much superior to Dee's, for he points out that the belligerent image is inverted; it will be erect only to a figure standing inside the focal point (between F and A in Figure 11), and he should see his opponent behind, not in front of, the mirror. Dee rightly calls his account of Pickering's glass "straunge to heare of."[66] The Elizabethan optical authorities knew that combinations of glasses had extraordinary powers. But they had not reduced them to effective rules, and nothing practical came from their knowledge, at least not directly.[67]

3. GEOCENTRIC ASTRONOMY

Ancient astronomy distinguished between two sets of reference circles, one fixed with respect to the observer, the other attached to the celestial sphere, which appears to revolve upon itself once in 24 hours. The chief circles of the first set are the meridian and the horizon, which comes in three types. The geometrical horizon is a circle tangent to the earth at the point of observation and extending to the sphere of the stars. Dee, following common usage, distinguishes further between the "true horizon," a plane parallel to the geometric horizon and passing through the earth's center, and the "sensible horizon," the finite conical surface described by the motion of a line drawn from the eye to the farthest perceivable points upon the earth's surface. As Dee observes (XLV), the extent of the sensible horizon depends on the height of the observer's eye above the ground.

In drawing the true horizon, no account is taken of terrestrial dimensions. For most practical astronomical purposes, the radius of the earth may be regarded as vanishingly small, and the geometrical and true horizons as coincident. For

66. "Preface," b.j^v; della Porta, *Magia*, lib. xvii.c.4; Schott, *Magia*, I: 327–328. It is, of course, easy to deceive oneself about the location of images in concave mirrors, especially if one has uninformed anticipations about where they should fall. Cf. V. Ronchi, *Optics*, tr. Rosen (1957), pp. 131–152.

67. Cf. de Waard, *Uitvinding*, pp. 69ff.

example (XLV, corollary), two fixed stars separated by less than a semicircle can be seen upon the (geometrical) horizon of only one place on earth at a given time, since only one qualifying true horizon exists, namely the plane containing the two stars and the earth's center. Stars diametrically opposite one another, however, appear upon the (geometrical) horizons of an infinite number of places simultaneously, for now an infinite number of planes can be drawn through the two stars and the center of the earth.

The second major fixed circle is the meridian, which stands perpendicular to the true or geometrical horizon and passes through the zenith, the nadir, and the north and south celestial poles. (The poles mark the intersection with the celestial sphere of the axis about which the sphere appears to turn.) The intersections of the meridian and the horizon (by which, when unqualified, either the true or the geometrical indifferently is meant) are the observer's north and south (Figure 12). A line through his position perpendicular to his

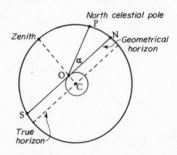

FIG. 12. Horizon and pole. S and N, south and north points of horizon; α, height of north pole; O, observer; C, earth's center.

north-south line gives him east and west. Note that the angular height of the pole above the horizon, α, just equals the latitude of the place, λ (Figure 13).

The major reference circles on the celestial sphere are the celestial equator and the ecliptic. The equator is the intersection with the sphere of a plane through the center of the earth perpendicular to the celestial axis (the line CP in Figure

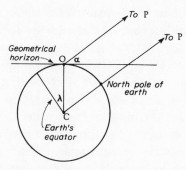

FIG. 13. Height of pole. Note that OP∥CP, whence angles α and λ have sides mutually perpendicular; O, P, C, and α have the same meanings they have in Figure 12.

13). This plane cuts the earth in the terrestrial equator, and the horizon at its east and west points. Every point on the celestial equator therefore rises due east and sets due west, and consequently spends equal time above and below the horizon. Dee and the older astronomers called this regularity "motion in the right sphere." Celestial points off the equator rise "obliquely," or "in an oblique sphere," never through the cardinal points, and spend more or less time above the horizon depending on their distances from the equator (XCIII). In particular, stars within an angular distance β of the pole never set (Figure 14). Note that the celestial equator rotates in place: although a different point of the equator

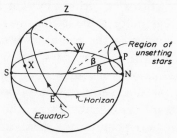

FIG. 14. Celestial equator, horizon, diurnal motion. X is a point on the sphere south of the equator; note that it rises south of east, sets south of west, and spends less than half its time above the horizon.

rises at each instant, the equator as a whole maintains the same position with respect to the horizon. This is not true of the ecliptic.

The ecliptic is the projection upon the sphere of the stars of the sun's apparent annual path. It is essential to distinguish this yearly circuit from the diurnal rotation of the stellar sphere, in which the sun—like every other celestial body or point—also participates. (The apparent diurnal rotation of the heavens from east to west, of course, arises from the earth's spin on its axis from west to east.) This participation causes day and night. But in addition to being carried by the world's motion from east to west, the sun, unlike the fixed stars, appears to move about 1° a day on a private path from west to east. In consequence of this slow drift against the motion of the stars the sun takes a little longer to return to the meridian —to rotate through 360°—than they do. Dee divides the time of the diurnal rotation, which he calls the equatorial period, into 24 "equal hours" (XLIV). The average time between the sun's successive crossings of the upper meridian (local noons) is about 4 minutes longer (cf. LXV); hence, by Dee's reckoning, a mean solar day lasts 24 and 1/15 equal hours. (Since we divide the mean solar day into 24 hours, our hours are about 10 seconds longer than Dee's.)

The ecliptic intersects the celestial equator at an angle of 23.5°. (This translates into Ptolemaic language the fact that the axis of the earth's spin inclines by that amount to the plane of its annual motion about the sun.) The points of intersection are known as the equinoxes because when the sun stands beneath the equator, days and nights are equal. At 90° from the equinoxes measured along the ecliptic lie the solstices, where the sun seems to stop before beginning its journey back toward the equator. One of the equinoxes, the vernal, is the reference point for the specification of all other positions on the celestial sphere. The angular distance of a celestial point measured eastward from the vernal equinox along the equator is its right ascension; its angular distance

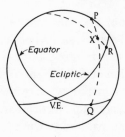

FIG. 15. Celestial coordinates. P is the pole of the equator (north celestial pole); VEQ is the right ascension of the celestial point or body X; QX is its declination; VER, its longitude; RX, its latitude.

north (or south) of the equator, its positive (or negative) declination. Distance measured eastward from the vernal equinox along the ecliptic is called longitude; north- or south-ward, latitude (Figure 15). The astronomers of Dee's day, in contrast to those of our own, measured from the ecliptic since the moon and the planets, the bodies of chief astrological interest, remain within a latitude of about 12°; whence the utility of defining a region of the heavens—the zodiac—straddling the ecliptic and containing the apparent planetary paths. By convention, each zodiacal sign occupies 30° along the ecliptic; the vernal equinox, summer solstice, autumnal equinox, and winter solstice sit at the first points of Aries, Cancer, Libra, and Capricorn, respectively (Figure 16).[68]

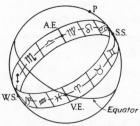

FIG. 16. The zodiac. P, A.E., V.E., W.S., S.S., the north celestial pole, autumnal and vernal equinoxes, winter and summer solstices.

68. For reference: Aries (Υ), Taurus (Θ), Gemini ($\mathrm{I\!I}$), Cancer ($\mathfrak{S}$), Leo (Ω), Virgo ($\mathfrak{M}$), Libra ($\triangleq$), Scorpio ($\mathfrak{M}$), Sagittarius ($\nearrow$), Capricorn ($\mathcal{V}_{\mathcal{J}}$), Aquarius ($\approx$), Pisces ($\mathcal{H}$).

The vernal equinox does not constantly occupy the same point on the celestial sphere, or rather remain at rest among the fixed stars. It appears to move along the ecliptic from east to west, in the direction of the diurnal motion and against the order of the zodiacal signs; or, what comes to the same thing, the vernal equinox may be taken at rest and the stars assumed to move away from it, parallel to the ecliptic and from west to east. The vernal equinox being the origin of coordinates, astronomers preferred the second alternative, and spoke of the "precession of the equinoxes," whereby they meant a revolution of the stars around the poles of the ecliptic in the order of the signs. (The phenomenon results from a wobble of the earth's axis arising from gravitational forces called into play by the earth's slight departure from sphericity.) Note that the "signs" therefore are not constellations but conventional coordinates; the inventory of stars in a given zodiacal sign changes continually. But not quickly. The precession amounts to about 49 seconds of arc a year, or one sign in 2200 years. Nonetheless, Dee recommends that we take it into account in exact astrology by distinguishing between the sidereal year, the time the sun takes to return to the same star, and the tropical or seasonal year, the period between the sun's successive visits to the same equinox (LXVI, LXVII).

Since the sun's light obliterates the stars', one deduces its position among them by reference to stars just visible before sunrise and after sunset, or to those crossing the upper, visible, meridian as the sun passes the lower, etc. Traditionally, astronomers distinguished nine such observational situations, each with two, three, or even four subspecies. Take the configuration "early morning east wind," when the star rises with the sun. One can distinguish the subspecies "visible morning fore-rising," "invisible morning after-rising," and "morning co-rising," according to whether the star just precedes, just follows, or accompanies the sun's crossing of the horizon. Ptolemy distinguished twenty-six subspecies in the nine observational situations,[69] to each of which Dee, with his characteristic concern to multiply possibilities, assigns an astrological importance (CVIII).

One more aspect of the sun's apparent motion needs attention. The picture so far presented suggests that the seasons have equal length, that the sun requires the same time to travel the 90° between successive equinoxes and solstices. In fact, because of the ellipticity of the earth's orbit summer is longer than winter. Rather than assign a varying speed to the sun, the ancient astronomers gave it a constant-velocity path eccentric to the earth (XLVII). In Figure 17, which repre-

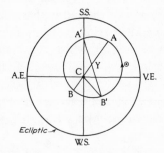

FIG. 17. Orbit of the sun. Y is the center of the eccentric; C (the observer), the center of the zodiac; the cardinal points as in Figure 16.

sents the plane of the ecliptic, the sun describes the circle AA'BB' centered on Y a distance CY (known as the eccentricity) from the common center C of the earth and the universe. The sun moves equably as seen from Y. As seen by the observer at (or near) C, however, it appears to move more quickly around perigee B than around apogee A; $\angle AYA'$ (= $\angle BYB'$), which measures the speed about A and B as seen from Y, exceeds $\angle ACA'$ and falls short of $\angle BCB'$ which measure, respectively, the speed around apogee and around perigee as seen from C. With appropriate orientation of the "line of apsides" BCYA in the zodiac, the sun's regular motion on its hypothetical

69. Ptolemy, *Almagest*, lib. viii.c.4.

"eccentric" AA'BB', when projected against the fixed stars by an observer at C, gives a fair approximation to the observed irregularity of the sun's apparent course along the ecliptic.

The moon's orbit inclines to the ecliptic at about 5.2°. The intersections of orbit and ecliptic are called "nodes." To account for the apparent motion of the moon, which (owing to the strong gravitational perturbing force of the sun) is much more complex than the sun's, the old astronomers gave the lunar orbit a changing eccentricity and made its nodes "regress"—move around the ecliptic against the order of signs—once in eighteen years. The moon circulates in its orbit from west to east, returning to the same star in 27.3 days (the "sidereal period") and to the same phase, say new moon, in 29.5 days (the "synodic period"). The difference is owing to the displacement of the sun, which completes about 27° in a lunar sidereal period; the moon needs an additional $(27.3) \cdot (27)/(360 - 27.3) = 2.2$ days to catch up (LXVIII). An eclipse can occur only when the moon arrives at or very near a node, that is, stands under the ecliptic, whence the name (Figure 18).

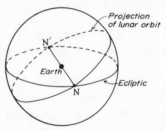

FIG. 18. Orbit of the moon. N is the "ascending" node (the moon there crosses into northern, or higher, latitudes), N' the "descending" node. Eclipses are possible only when earth, moon, and sun are colinear, i.e., when the moon is at a node.

Despite its complexity, the motion of the moon is in one respect very much simpler than that of any planet: like the sun, the moon always runs along its private orbit in one direction, from west to east. The planets, however, although they usually move in their own orbits in the order of the signs, also periodically appear to slow down, stop, move toward the west ("retrograde"), and stop again before resuming their journey to the east. The old astronomers represented this behavior (which arises from the annual motion of the earth) by mounting the planets on auxiliary circles or "epicycles" (LXII, LXX), the centers of which revolved about an eccentric (which approximates the elliptical paths of the planets) in the manner described for the sun (Figure 19).[70] The center of the epicycle completes its

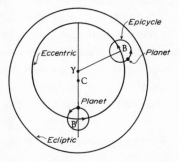

FIG. 19. Eccentric and epicycle. C is earth; Y, center of the eccentric; B, B', center of the epicycle; CY = ϵ = eccentricity. Note that the planet revolves in its epicycle in the same direction (counterclockwise) as the center of the epicycle rotates on the eccentric.

revolution in the planet's sidereal period; the planet revolves once about the epicycle center in a planetary synodic period, the interval between retrogradations.

Retrogradation occurs only at or near "opposition," when the planet stands in the opposite direction from the sun as

70. The refined Ptolemaic system referred the regular rotation of B not to Y but to a point called the "equant" lying on the line of apsides in the same direction as Y from C, but at twice the distance (CY = Y · Equant). Regulation by the equant gives an excellent approximation to motion in a Kepler ellipse of small eccentricity. Since Dee does not mention this fine device, however, we shall ignore it.

seen from earth; and opposition occurs only when the center of the planet's epicycle comes to perigee (B′ in Figure 19). If one arranges the directions of rotation of epicycle and eccentric properly, then, when the planet comes near B′C, its velocity perpendicular to the line of sight deriving from its rotation in its epicycle will oppose the velocity it obtains from the revolution of the epicycle around the eccentric. For appropriate relative dimensions and speeds of rotation of epicycle and eccentric, the velocities transverse to the line of sight can come equal. The planet will then appear to stop. Should the epicyclic velocity from east to west exceed the eccentric velocity from west to east, the planet will retrograde. A further complexity in planetary motions, one which we have already considered in the case of the moon, namely motion in latitude (XVII), can be assimilated by inclining the plane of the epicycle to that of the eccentric.

Planetary sizes and distances

The ancient astronomers invented a procedure excellent in theory but hazardous in practice for finding the distances of sun and moon from earth in terms of terrestrial radii. First, one obtains the solar distance, d_s, in terms of the lunar, d_m, by measuring the angle α between the centers of sun and moon at the precise instant of half-moon, when angle SME (Figure 20) is 90°. Evidently $d_s/d_m = \sec \alpha$. The ancient and

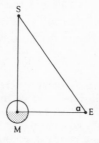

FIG. 20.   Relative distances of sun and moon: $d_m = \text{ME}$, $d_s = \text{SE}$.

medieval astronomers underestimated a; and since the secant changes rapidly in the region of a, namely 90°, they made the ratio of distances too small, about 20 rather than 400. (The figures refer to mean values: owing to the eccentricities of the orbits the perigee of the moon is about $(5/6)d_m$.)

The reduction of these figures to t.r. (terrestrial radii) may be made with the help of a lunar eclipse. In Figure 21,

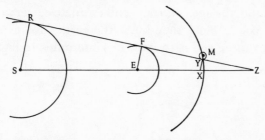

FIG. 21. Absolute distances of sun and moon. RS ($=s$), EF ($=e$), are the radii of sun and earth, and XY∥EF∥RS approximate the length of the moon's course through the earth's shadow.

the moon M is about to enter the shadow cast by the earth E, and the distance XY∥EF∥RS approximates half the path of the eclipsed moon. Let e, m, s be the radii of earth, moon, and sun, respectively. From the similar triangles ZRS, ZFE, and ZYX,

$$\frac{e}{s} = \frac{d_m + x}{d_m + d_s + x} \quad \text{and} \quad \frac{e}{\text{MX}} = \frac{d_m + x}{x},$$

where $x =$ XZ. These equations can be made to yield a value for d_m as follows. Measure the time for the moon entirely to immerse herself in the shadow, say t_1, and the time between the beginning of immersion and that of first appearance, t_2. Then MX $= (t_2/t_1)m$.[71] Furthermore, the angles subtended by sun and moon at the earth are about equal; one easily

71. 2MX $= vt_2$, and $2m = vt_1$, where v is the moon's velocity. For details see Duhem, *Système*, II: 1–34.

determines the value ($\beta \simeq 0.5°$), and hence the ratios $m/d_m = s/d_s = \tan \beta$. Introducing the known quantities into the equations and eliminating x, one finds

$$d_m = \frac{1 + \cos \alpha}{1 + t_2/t_1} \cdot \text{ctn} \, \beta \text{ t.r.}$$

Since $\cos \alpha \ll 1$, d_m does not depend sensitively upon it, or what is the same thing, d_m/d_s. Hence, despite their poor value for that ratio, the ancients could still estimate d_m very closely, at about 60 t.r. Since they knew the earth's radius e, they could convert d_m and d_s ($\simeq$ 1200 t.r.) into miles. In Figure 22,

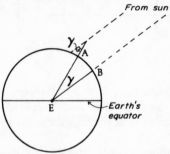

FIG. 22. Measuring the earth's radius. A and B are stations on the same meridian at a distance p apart; γ is the angle of the shadow cast by a vertical pole at A when the sun is in the zenith of B. Since $\angle AEB = \gamma$, $2\pi e : p :: 360° : \gamma$.

an upright stick at some place A casts a shadow of angular width γ at the same moment that, at another place B on the same meridian, the sun stands directly overhead. From the measurable angle γ, which is also the difference in latitude between the places, and from the distance AB walked off in miles, one easily obtains e with the help of the golden rule: $\gamma : 360° :: AB : 2\pi e$.

The distances of the planets and fixed stars cannot be obtained from Ptolemaic devices—indeed, even the order of the planets cannot be established—without two special assumptions. The first sets the order of planets by placing the slowest-moving—those that fall furthest behind the diurnal motion of the heavens—closest to the earth. The principle yields the sequence moon, Mercury, Venus, sun, Mars, Jupiter, Saturn. The second assumption is that there is no waste space anywhere in the universe. The apogee of the sun, for example, must lie at the same distance as the perigee of Mars, the apogee of which determines the perigee of Jupiter; carrying the procedure to its logical conclusion, the ancients located Saturn's apogee in the sphere of the fixed stars. Since one knew the absolute mean distance of the sun and the Ptolemaic devices established the ratios of eccentricity and of the epicyclic radius to the radius of the eccentric, one could compute all the planetary distances. In particular, Mars's perigee lies at $r_P = d_s(1 + \epsilon_s)$, ϵ_s being the eccentricity of the solar orbit; and Mars's apogee lies at

$$r_A = r_P(1 + \epsilon_{Mars} + \eta)/(1 - \epsilon_{Mars} - \eta),$$

where η is the ratio of Mars's epicyclic to its eccentric radius. Proceeding in this manner, all astronomers from Ptolemy to Dee made the radius of the sphere of the fixed stars about 20,000 t.r.[72] Part of the reason for the longevity of the doctrine of plenitude was the extraordinary coincidence that the mean solar distance d_s computed from the mean lunar distance d_m using the Ptolemaic models of the moon, Mercury, and Venus agreed very well indeed with the entirely erroneous value of d_s—1200 t.r.—obtained from direct observation of the half-moon.

In marked contrast to Ptolemy's system, Copernicus's can be made to yield both the order and distance of the planets without additional assumptions. For Mercury and

72. Dee gives 20,081.5 t.r. of 3436 and 4/11 miles each in "Preface," b.ii^r; Bacon makes it 20,100 in *Op. maius*, ed. Bridges, I: 227 (Burke, I: 249). For details of the traditional calculation see W. Hartner in *Mélanges A. Koyré* (1964), II: 254–282; B. R. Goldstein in Amer. Phil. Soc., *Trans.* 57 (1967): 3–55; F. S. Benjamin, Jr. and G. J. Toomer, *Campanus of Novara* (1971), pp. 327ff; and J. Henderson in *Copernican Achievement*, ed. Westman (1975), pp. 113–127.

Venus, one computes mean distances in terms of the radius g of the earth's orbit merely by measuring ϕ, the planet's maximum elongation from the sun. Since, in this case, $\angle SPE$ (Figure 23) is 90°, $SP = r$, the planet's mean solar distance,

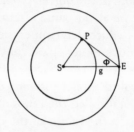

FIG. 23. Copernican determination of the solar distances of the inferior planets. P, E, S, the planet, earth, and sun, respectively; g ($=d_s$), the mean radius of the earth's orbit.

equals $g \sin \phi$. The mean distance of the planet from earth is $(1/2) \cdot (g - r + g + r) = g$. For the superior planets, the calculation of mean solar distance requires the measurement of θ, the angle between sun and planet at a time t after opposition (Figure 24). If the mean sidereal periods of earth

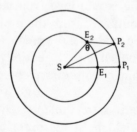

FIG. 24. Copernican determination of the solar distances of the superior planets; E_1, E_2, P_1, P_2, successive positions of earth and planets; $SE_1 = SE_2 = g$.

and planet are τ_E and τ_P, $\angle E_2SP_2$ gives SP via the proportion

$$SP_2 : g :: \sin \theta : \sin \left[\theta + 2\pi \left(\frac{t}{\tau_E} - \frac{t}{\tau_P} \right) \right].$$

The mean distance of the planet from earth is $(1/2) (r - g + r + g) = r$. The Copernican dimensions for the outer planets fall short of the Ptolemaic by a factor of two or three:

Mean Values of the Planetary Distances from Earth
(in 1000 t.r.)

	Campanus[73]	Alfraganus[74]	Copernicus[75]
Mars	5.6	5.05	1.7
Jupiter	13.1	11.6	6.0
Saturn	19.4	17.3	11.0

Knowing the distances r of the celestial bodies in t.r., one can compute their diameters $2a$ from the angles ϕ they subtend at the earth. Since the angles are all very small, $2a = \phi r$ in radian measure (1 radian $= 57°$). Now it is very difficult

Stellar Sizes (a^3/e^3)

Class of Star[76]	Alfraganus[77]	Dee	Brahe[78]
I	107	107	68.0
II	90	80	28.5
III	72	70	11.0
IV	54	54	3.4
V	36	30	1.0
VI	18	18	0.33

73. Benjamin and Toomer, *Campanus*, 326–343.

74. Alfraganus, *Libro*, ed. Campani (1910), p. 146, repeated by Bacon, *Op. maius*, ed. Bridges, I: 227–228 (Burke, I: 249–250).

75. Copernicus, *De Rev.*, lib. v.c.14, 19, 21, using the mean value of solar distance, 1142 t.r., from Ptolemy, *Almagest*, lib. iv.c.21.

76. The fixed stars are divided into six classes by apparent brightness, the sixth class comprising those just visible to the naked eye. Ptolemy, *Almagest*, lib. vii.c.5 and viii.c.1.

77. Alfraganus, *Libro*, pp. 149–150, retailed by Bacon, *Op. maius*, ed. Bridges, I: 235–236 (Burke, I: 257–258). For variants see Goldstein in Am. Phil. Soc., *Trans.* 57 (1967): 11; J. L. E. Dreyer, *History of Astronomy*, 2nd ed. (1953), p. 258; and Duhem, *Système*, II: 52–53.

78. Brahe, *Opera omnia*, II: 430–431. Brahe also repeats Alfraganus's values (*ibid.*, pp. 417–418), which he rightly criticizes as arbitrary.

without a telescope to measure ϕ for the smaller planets, and impossible to do so for the stars; yet the Arab astronomers supplied numbers which, with trifling variations, had become standard by the twelfth century, when the technical treatises of the East began to come West. The values for the "magnitudes" of the stars relative to the earth's (i.e., relative volumes, a^3/e^3) that Dee gives (LXXVIII) may be compared to the traditional figures of Alfraganus and the innovations of Tycho Brahe as shown on the previous page.

The figure 107 apparently came from taking stars of the first magnitude to be somewhat larger than Saturn (4.75 : 4.50); Alfraganus then reckoned the others by dividing 106 into six equal parts. The discrepancies between Dee and Alfraganus may have arisen from corruptions in Dee's manuscript of his Arab original, or from sloppiness or enterprise of his own.[79]

4. ASTROLOGY

Astronomical and physical principles serve Dee's astrology chiefly by making possible a computation of the strength of the rays or species emitted by celestial bodies at diverse times and places. This strength, as received at a given terrestrial point, depends upon parameters of motion, such as the distances of celestial bodies from earth (XVII) and the duration of their passage above the horizon (LV, LVI); and upon physical interactions among the planets deriving from their changing aspects, especially from multiple conjunctions (CXV–CXVIII).

The computation of planetary distance requires knowledge of the precise position of the body in its eccentric and epicycle, including its excursions, if any, in latitude (LXXXIX).

79. Cf. James, *List*, p. 12, no. 26. Cf. Dee in Euclid, *Elements*, ed. Billingsley, f. 389^v: with the theorems of Archimedes *et al.* the "spherical soliditie" of the heavenly bodies may "with as much ease and certainety be determined of, as of the quantities of any bowle, ball or bullet, which we may gripe in our handes."

We are also to consider that the distance of a celestial body from a terrestrial point changes during the day even if the body does not move against the fixed stars (XLVII). The ratio of the greatest distance, when the body rises or sets, to the least, when it crosses the meridian (Figure 25), is

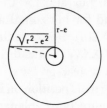

FIG. 25. Diurnal change in the distance of a celestial body, distant r from the center of the earth; e is the earth's radius.

$\sqrt{(r + e)/(r - e)}$. In the case of the moon at mean distance, $r = d_m = 60$ t.r., the ratio amounts to 1.017; for the fixed stars it is 1.00005.

The duration of passage above the horizon, or *mora*, depends upon the planet's position in its path (LXXXVII). All planets in direct motion, and the sun and moon always, spend longer above the horizon than do the celestial points or stars with which they rise; for their private motions carry them to the east while the rotation of the heavens hurries them west. The easting made good determines the excess of the planetary, solar, or lunar *mora* over that of co-rising celestial points. When retrograde, planets complete their paths more quickly than the co-rising points, their *morae* are shorter and their influence diminished (LXXXVIII). Since the apparent speeds of planets, sun, and moon constantly alter, the time any of them takes to return to the meridian (the diurnal period, LXIX) might well differ from what Dee calls the "horizontal period" (LXXXII), the interval between successive risings.

The same point may be made by reference to the stars. In a natural day (the interval between successive noons) the stars must rotate somewhat more than 360° to make good the sun's easterly drift. The excess beyond 360° is known as the "diurnal direction" of the stars (LXXIX); its magnitude depends upon the sun's right ascension, being greatest at the solstices, when his motion along the ecliptic parallels the equator (LXXX). Dee directs the astrologer to compute directional progress not only for the sun, but also for the moon (LXXIX) and planets (LXXXIV), and for horizontal as well as for diurnal periods (LXXXII, LXXXIII). Retrograding planets have "reversed directions" (LXXXV). The directions of a superior planet summed over a year give its "annual progression," which varies according to its position on its eccentric (LXXXVI).

Like *morae* and directions, planetary aspects may be computed from the standard astronomy; only in interpreting their significance is an astrologer wanted. Aspects are angular separations of celestial bodies as seen from earth. Astrologers assigned special power to five angles or, rather, angular separations within some conventional intervals around the special angles: 0°, or conjunction; 60°, or sextile; 90°, or quartile; 120°, or trine; and 180°, or opposition.[80] The Arabs had associated conjunctions with great changes in human history,[81] and Dee attends only to conjunctions among the aspects; but he approaches them not as an Eastern world historian, but as a Western pedant somewhat leavened by cabalistic mathematics.[82]

How many different sorts of conjunctions are there? One must distinguish. Ignoring the fact that the planets, moon, and sun have different powers, Dee finds (CXVI) 21 different conjunctions possible involving two and only two of the seven bodies; 35 involving three; and, in general, $7!/(7 - n)! \, n!$,

80. Cf. A. Bouché-Leclerc, *L'Astrologie grecque* (1899), pp. 165–179. In practice, any separation within, say, 10° of the privileged angles counts.
81. Cf. F. Bold, C. Bezold, W. Gundel, *Sternglaube*, 5th ed. (1966), pp. 34, 110–111, 135.
82. Cf. M. Turetsky, *Math. Teacher* 16 (1923): 29–34. All the combinatorial mathematics Dee required may be found in Levi ben Gerson's *Sefer ma'aseh hosheb* (1321) or *Die Praxis des Rechners*, tr. Lange (1909), pp. 47–55, 84–85.

where n is the number of conjunct bodies and the symbol $n!$ signifies the product of all integers from 1 to and including n. The total of the different possible sorts of conjunctions for the planets, irrespective of relative power, is 120. Such calculations offered no difficulty to Renaissance mathematicians. Indeed, they were possible in the Middle Ages; the 120 conjunctions already figure in Rabbi ben Ezra (c. 1140), from whom Dee's analysis perhaps ultimately descends.[83]

Admitting now that the seven bodies may have different powers, Dee distinguishes two cases: (1) no two powers are the same (CXVI) and (2) the powers can be equal or unequal (CXVII). In Case (1), the number of different n-body conjunctions may be obtained by multiplying the earlier formula by $n!$: each two-body conjunction, for example, now can occur in two ways, with A stronger than B or vice versa, giving 42 possibilities in all. For three bodies we have $7!/4! = 210$, and for all seven, $7! = 5040$. In Case (2), two or more planets may have equal strength. If two only are equal, the pair may be chosen from the seven in $7!/5! \, 2! = 21$ ways; the relative vigor of the pair and the five remaining bodies may be ordered in $6!$ ways, making, for the case of two and only two equal bodies, $21 \cdot 720 = 15{,}120$ possibilities in all. In general, one multiplies the number of n-body conjunctions of the first case, when no account is taken of relative strength, by $(8 - n)!$ The total is 20,301, to which Dee adds the 5040 conjunctions when no two planets have equal strength, making in all 25,341 different possible conjunctions.[84]

How is this vast mathematical apparatus—these *morae*,

83. J. Ginsburg, *Math. Teacher* 15 (1922): 347–356. The problem of permutations, discussed in reference to the number of different arrangements of people sitting about a table, appears in the popular *Summa* of L. Pacioli (1494). See D. E. Smith, *History of Mathematics*, 2nd ed. (1958), II: 524–529. Such problems also interested Cardano.

84. Dee slipped in computing the number of conjunctions when five and only five bodies are equal: the five can be chosen in $7!/5! \, 2! = 21$ ways, and can be ordered among the remaining two in $3! = 6$ ways, making 126 (not 120) possibilities.

directions, distances, and periods, not to mention the 25,341 conjunctions—to be worked by the astrologer? How, if he lives long enough to make the calculations, can he compute celestial influences upon the "entire and unchangeable order of every generation" (XXI); upon events both immediate and proximate (LVII); upon the "chief and truly physical causes of the procreation and preservation of all things which are born and live" (CVII); upon the cause and occasion of the death of the body (CIX)? Dee's counsel in these matters is not very helpful.

The species of each star and planet are characteristic of it (L, LXXIV, XC); astrological influence takes place (as we know) via a kind of species which obeys the laws of reflection and refraction, but which penetrates more deeply into matter than the light that usually accompanies them (XXV). In a word, the astrological species resemble the magnetic (XXIV), with the important exception that magnetism has only local significance, whereas the heavenly bodies "move"[85] everything here below (XVI, XVII, CXIV). And what are the characteristics of each celestial astrological species? Dee identifies only the sun as the principle of heat, and the moon as that of moisture (CII–CIV), ancient and obvious associations.[86] Both bodies act more powerfully the closer they are to earth; and the moon's power over moist things also increases with her apparent velocity (CIV, CV). For the rest, Dee hints that no star or planet is evil in itself, but may do harm when irradiating "corrupt matter" (CXII, VII). This Renaissance optimism, which we have also found in Mercator, contrasts with the traditional interpretation of certain celestial bodies, particularly Mars and Saturn, as inveterate evildoers.[87] The same

85. "Motion" here has its Aristotelian sense of change in general (*Physica*, 201ª10–15).

86. Cf. Bouché-Leclerc, *Astrologie grecque*, pp. 89–93.

87. Mercator, see chap II: 1, above; Bouché-Leclerc, *Astrologie grecque*, pp. 93–99; Ptolemy, *Tetrabiblos*, tr. Robbins (1940), p. 39; Bacon, *Op. maius*, ed. Bridges, I: 142–143.

optimism may be seen in Dee's assertion (XC) that a planet need not cease to be beneficent when "combust," or close to the sun. This again ran counter to an old astrological teaching, that the sun's heat injured the rays from bodies in conjunction with it.[88]

To proceed further, the astrologer must find by experience (LXXIII) the special bonds between particular elemental bodies and individual celestial ones, and the peculiar force of each planetary conjunction.[89] The research will not be easy. For one, the earth continually bathes in astrological rays (XCI) whose make-up and strength continually alter; every point in the universe at every instant enjoys a unique radiation (LI), whence the great diversity in sublunary affairs (LXXVI, CXIII). Nor are the complexity and variability of the radiation received on earth the only obstacles to a correct evaluation of the virtue of each celestial body. One must also consider that the effects of the rays depend upon the nature of the receiver (VII, XXVI). The sun simultaneously melts wax and dries mud: should we ascribe to it a drying or a wetting power?

Dee does not abandon us altogether. In unscrambling the various celestial influences, we should bear in mind the old principle that both similarity and difference must exist between agent and patient (VIII, IX). It will also help to consider the very great harmony among all created things, and especially analogies between the celestial, elemental, and microcosmic regions. Take, for example, the illuminating correspondences between the sun, gold, and the heart of a man (LXXIII). But we are not limited to these general principles. By art, by our own manipulations, we can separate, strengthen, and concentrate particular celestial influences in appropriate receivers (XXVII), thereby simultaneously affording

88. Bouché-Leclerc, *Astrologie grecque*, pp. 112, 309.
89. Cf. Bacon, *Op. maius*, ed. Bridges, II: 215–216 (Burke, II: 627–628), recommending that astrology be developed as a *scientia experimentalis*, or, to use Dee's special jargon, as a branch of Archemastrie.

ourselves opportunity for study and for making useful talismans, including the philosopher's stone (LXXVII). Here, catoptrics will be especially helpful, both in imprinting rays via burning mirrors (LII) and in teaching how to derive information about lunar powers by comparing the effects of a full and an eclipsed moon (LIII). Pyronomia (II), the science of fire and heat, plays an equal part. It directs the adept to observe, say, the heating produced by the sun at different times (XCVI–XCIX), and to compare its effects with those of the planets (C). The business can be reduced to number by the art of graduation (XIX, XX), the construction of an arbitrary scale of "degrees of heat".[90]

Still we are far from bringing the motions of the heavens into the affairs of men. To do so requires methods beyond the physical astrology with which Dee tries to make do. He hints at two of these methods, the significator (LXXXIX) and the houses (LXIII), both of which arbitrarily couple specific celestial bodies with particular stages in life. In the first method, one predicts, say, the circumstances of marriage by examining the situation of the relevant significator—usually the moon for a man and the sun for a woman—in the sky at the time of birth.[91] The practiced astrologer will consider the position of the significator with respect to the horizon, its location in the zodiac, its angular distance from other planets, and, if he follows Dee's injunctions, its distance from earth (LXXXIX).

The house system relates twelve general areas of human concern to the zodiac according to its position with respect to the horizon at the moment of birth. The traditional areas are (1) nature of the newborn, (2) his fortune, (3) siblings, (4) parents, (5) children, (6) illnesses, (7) marriage, (8) death, (9) travels, (10) distinctions, (11) friends and good fortune, (12)

90. Chap. I: 2, above.
91. Bouché-Leclerc, *Astrologie grecque*, pp. 447–450; Ptolemy, *Tetrabiblos*, pp. 393–409.

enemies and bad fortune (Figure 26). This scheme is associated with the zodiac by taking the line HO to represent the horizon, H being the "horoscope" or zodiacal rising point at the moment of birth, and MC (medium caeli)-IMC (imum medium caeli) to represent the meridian. One places the

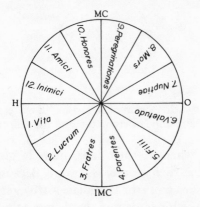

FIG. 26. Standard house scheme. H, O, MC, IMC, the horoscope, occidens (setting point of zodiac), medium caeli (midheaven), imum medium caeli (lower midheaven), respectively.

ecliptic points H and O, respectively, at the beginnings, or cusps, of the first and seventh houses. How much of the zodiac then falls into each house?

Dee mentions three ways of cutting the pie. The simplest, that of equal houses, carves the zodiac into 30° pieces; if the horoscope lies, say, at 0°   , the cusp of the twelfth house is at 0°   , that of the eleventh at 0°   , and that of the tenth, 90° from H and the highest point of the ecliptic above the horizon, at 0°   . This is the system that Bacon preferred.[92] It has the disadvantage that the MC, which Dee calls "cor caeli" (XCIII), does not in general coincide with the cusp of

92. *Op. maius*, ed. Bridges, I: 254–258 (Burke, I: 277–280). On the problem of house division see W. A. Koch and F. Zanziger, *Regiomontanus* (1960).

the tenth house; for, as appears from Figure 27, the meridian bisects the half of the ecliptic above the horizon only in two special cases, only when an equinox (0°♈ or 0°♎) is rising. Now celestial bodies are supposed to exert their greatest

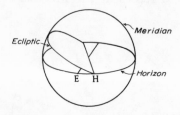

FIG. 27.   Ecliptic with a summer sign rising. E the east point of the horizon; H the horoscope or rising point of the ecliptic.

power on us when at their maximum height above the horizon, when at their prime, as it were. Because they attain this height as they cross the meridian, astrologers located the tenth house, which determines the degree to which one can rise in life, in the region just preceding the meridian.[93] Since the equal-house system allows such blemishes as the MC in the ninth house, many scrupulous astrologers refused to have anything to do with it.

Dee notices two alternatives (LXIII), which agree in fixing the cusps of the tenth and fourth houses at the MC and IMC, respectively, but which differ in placing the off-angle houses, 2, 3, 5, 6, etc. One, generally known as the system of Campanus, operates with the "prime vertical," the great circle that passes through the zenith and the east and west points of the horizon. The astrologer divides the prime

93. The significance attached to the other houses also derives at least in part from powers ascribed to the rising and falling of the heavenly bodies. Note the house of death at the descendent. The association of marriage with the Western horizon comes from another scheme in which not 180°, but 360°, is the analogy to life's course; hence setting occurs at the halfway point, and so might mark marriage. See Bouché-Leclerc, *Astrologie grecque*, pp. 256–288.

vertical into twelve equal portions beginning at the east point, and passes great circles through the points of division and the north and south points of the horizon; the intersections of the ecliptic with these circles fix all the cusps (Figure 28). This is

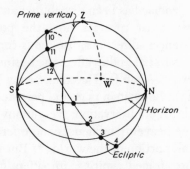

FIG. 28. Campanus's system of house division. The numbers give the cusps of the houses.

probably the system Dee intends by "the method of horizon circles." The alternative, the system of Alchabitius (or, in Dee's terminology, "of meridian circles"), takes the equator as basis, and meridians through the north and south poles of the heavens as the auxiliary circles whose intersections with the ecliptic determine the cusps (Figure 29). As Dee observes,

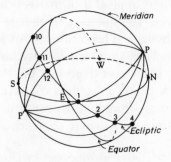

FIG. 29. Alchabitius's system of house division. A related scheme, Regiomontanus's, uses auxilliary circles that divide the equator and pass through the north and south points of the horizon.

the two systems become one at the terrestrial equator, for there the celestial equator coincides with the prime vertical and the poles with the north and south points of the horizon. His statement that only two systems remain at the poles appears to be an error.

Once the zodiac has been distributed among the houses, the astrologer must consider how the positions, aspects, velocities, distances, directions, luminous cones, virtues, and species of the celestial bodies affect the business of the houses in which they sit. A job for Hercules, or for IBM. Dee began to collect data at Louvain in 1548, and persisted for some years in "thousands . . . of observations (very many to the hour and minute) of the heavenly influences and operations actuall in this elementall portion of the world."[94] But the undertaking must have discouraged a spirit so impatient for knowledge as Dee's: for, to mention only the smallest matter, his observations could have acquainted him with the peculiar powers of but a fraction of his 25,000 different conjunctions. The many nativities he drew up in the 1560s are quite traditional.[95] Yet we must allow that Dee's method for studying the influences of the heavens failed not because it was frivolous, but because he used it where it did not work. The idea of a quantitative physical science in which experiment and mathematics continually reinforce and correct one another prospered greatly when others applied it to matters closer to home.

There remains the question of the tie between Dee's mathematical physics and his occultism. An adequate answer would require more biographical and psychological information than we are likely ever to possess. But the general trend of his life's work makes possible a conjecture. Each major shift in Dee's studies—from his early concerns to alchemy and the *Monas*, and from the *Monas* to the fanciful geography and

94. *AT*, pp. 5, 28. Other nativities cast by Dee are alluded to in *Private Diary*, ed. Halliwell (1842), pp. 1–2. Cf. Clulee, *Glas*, pp. 118–120.

95. Clulee, *Glas*, p. 120, referring to Ashmole MS 337, ff. 20–57 (Bodleian).

angelic conversations—took him farther and farther from applied mathematics. It appears that, for him, doing mathematics and practicing magic were antithetical. We may guess that in this respect he was not unusual. Consider Agrippa's estimate of the productions of the technical astronomers of his time. "Their vain disputes about Eccentricks, Concentricks, Epicycles, Retrogradations, Trepidations, accessus, recessus, swift motions and circles of motion," says he, "[are] the works neither of God nor Nature, but the Fiddle-Faddle and Trifles of Mathematicians."[96]

96. Agrippa, *Vanity of Arts and Sciences* (1676), p. 86. The first Latin edition was published about 1530.

Propædeumata Aphoristica

IOANNIS DEE, LONDINENSIS,
De Præstantioribus quibusdam
Naturæ Virtutibus.

SVPERCAELESTES   ET TERRA FRVCTVM
RORETIS AQVAE.   DABIT SVVM.

QVATER △ NARIVS
IN TERNARIO CONQVIESCENS.

Londini.
Anno. M.D.LXVIII.

AN APHORISTIC INTRODUCTION
By John Dee of London
Concerning Certain Outstanding
Virtues of Nature

LET THE

WATERS ABOVE THE   AND THE EARTH WILL

HEAVENS FALL   YIELD ITS FRUIT

THE QUATERNARY

RESTING IN THE TERNARY

London

1568

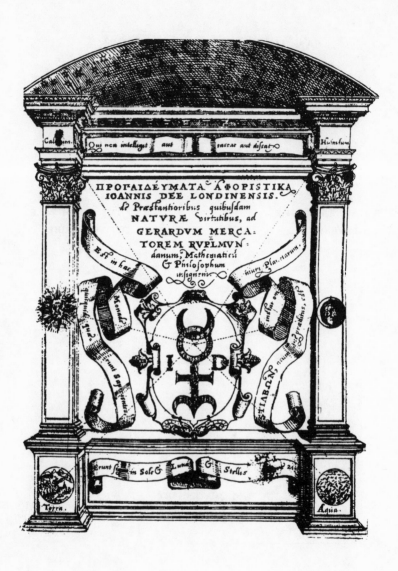

Hot   Moist

Let him who does not understand either be
silent or learn

AN APHORISTIC INTRODUCTION
By John Dee of London
Concerning Certain Outstanding Virtues of Nature
To Gerard Mercator of Rupelmonde
Distinguished Mathematician and Philosopher

In   MERCURY,
    this   endowed with
        Monad   a sting,
                is   is like
whatever   all
    wise   the
        men   planets
            seek

And there shall be signs in the sun, and in
the moon, and in the stars—Luke 21

Earth   Water

✺ *PROPAEDEUMATA APHORISTICA*

Translated by Wayne Shumaker

ACKNOWLEDGMENTS

Among the persons whose help on details has not slipped my memory, I wish to thank the following: my colleagues John Coolidge and Joseph Fontenrose; Jonathan Tuck, a graduate student in English; William J. Fulco, S.J., of the Jesuit School of Theology, Berkeley; Edward C. Hobbs, of the Church Divinity School of the Pacific, Berkeley; and David Winston, of the Graduate Theological Union, Berkeley.

W. S.

≈ Note on the Text

The basic Latin text is that of 1568, printed at London by
Reginald Wolfe (d. 1573), Latin Printer to the King. I have not
seen the book itself but have worked from a print of the
University of Michigan microfilm. The only other known
edition is that of 1558, printed in London by Henry Sutton.
This, too, I have seen only in a microfilm print-off. Although
I first made a complete typescript of the earlier text, having
been misled by a statement in the 1568 edition that the later
printing was "unchanged in ... number, order, and substance" (*Lectori Philosophiae sincerioris studioso*), the second
edition clearly supersedes the first. Not only are we told that
"The edition issued in 1558 limped in many places from the
great carelessness of the printer" (*Typographus Lectori*), but a
considerable number of verbal changes has been made,
expansions appear, and "Corollaries" and "Inferences" have
been added. The printing is also better, especially in the
Greek excerpts (where the font is more modern and abbrevi-
ated forms are less common), and the punctuation is improved.

Nevertheless, the textual notes record all the substantive
differences between 1568 and 1558. Differences merely in
spelling, punctuation, capitalization, and italics are not
noticed. The notation "*1558 omits*," when unaccompanied by
an explanation, has to do only with the single word preceding

the reference number. A note like "*1558* ipsis" means that a single word has been changed. For the rest, I trust that the textual notes will be self-explanatory.

The simplest explanation of my general editorial practice is that the text has been normalized, but only within limits. The tittle has become an *m* or *n*; ę has been expanded to *ae*; *foelix* has become *felix, coelum* has become *caelum*, and *ocium* has become *otium. Caetera* and *pene* (for *cetera* and *paene,* obviously) have, however, been preserved because they are alternative dictionary forms, though unapproved, and *seculum* has not been expanded into *saeculum* or *ex-* into *exs-* (as in *exstare, exsistit*) because of a scruple—no doubt ridiculous— against the supplying of a letter wholly missing in the original. On the other side again, *u*'s and *v*'s have been normalized, and *-ij* has become *-ii.* The capitalization, punctuation, and italics of the original have been preserved without change, as has also (in the Latin text but not in the English) an inconsistent use of Roman and Arabic numerals. The Latin accent marks common in Renaissance texts have been omitted; the Greek accents, although not always correct, have sometimes been added but never altered. No attempt has been made to discover the sources of Greek quotations. Quotation marks regularly indicate Greek phrases and sentences.

Typographus Lectori.

Habes hic Candide Lector, hanc secundam istorum A-
phorismorum editionem,[1] *longe emendatissimam, ex ipsius*
Authoris autographo, accuratissime impressam. Illa enim
quae Anno 1558 emissa erat, magna Typographi in-
curia, permultis claudicabat locis, veluti tu ipse,
ex diligenti nostrorum laborum collatione,
facillime iudicare possis. His igitur
utaris, fruarisque. Vale.
Anno a partu Virgineo 1567
Mense Decembri, Londini.

1. aeditionem

The Printer to the Reader.

You have here, Candid Reader, this second edition of those
Aphorisms, much emended and printed with great accuracy
from the autograph of the author himself. The edition
issued in 1558 limped in many places from the
great carelessness of the printer, as you yourself
may judge easily from a diligent comparison
of our labors. Use this, therefore,
and enjoy it. Farewell.

In the year 1567 from the Virgin birth.
In the month of December, London.

Clarissimo viro D. Gerardo
MERCATORI, RUPELMUNDANO,
Philosopho & Mathematico
illustri, ac amico suo lon-
ge charissimo,
IOANNES DEE, LONDINENSIS,
S. D. P.[2]

UNdecimus iam agitur annus (humanissime, doctissi-
meque mi Gerarde) ab illo, quo nostris ego relictis Academiis,
omnibusque nostrarum scholarum, in artium septem
(liberalium dictarum) professione, percursis ordinibus: sine
subere (ut in proverbio est) nare, & in Regiones transmarinas
coeperam peregrinari, ad ipsos investigandos fontes, a quibus
hac nostra aetate, plurimi ad nos optimarum quarumque
Artium deducebantur canaliculi: & cum illis vitam ducere
familiarem, quorum vel levissimus quisque unius diei in
scribendo, labor, nobis antea domi desidentibus, per anni fere
unius spatium, satis (ad intelligendum) faceret negotii. Atque
in isto primae meae peregrinationis inchoato cursu, quoniam
in te, primum omnium, Lovanii tum agentem, incidere,
maximo mihi summi Numinis obtigit favore: & ex tuis
mecum disceptationibus, tum primas tum altissimas ut
radices ageret tota mea peregrina philosophandi ratio: Nunc
proinde ego esse aequum censeo, rationique maxime consen-
taneum, ut iam primo peregrinantes, laborum etiam tu
meorum primitias, iure tibi vendices meritissimo. Et maxime,
cum mutuae nostrae amicitiae, familiaritatisque consuetudo
ea erat, toto ut triennio, vix totos tres simul dies, alter
alterius lubens careret aspectu: & ea utriusque nostrum
discendi, philosophandique aviditas, ut postquam convenire-
mus, tribus vix horae minutis, ab arduarum & utilissimarum
rerum indagatione abstineremus. An non huius nostrae tam
sincerae amicitiae, & tam suaviter continuatae philosophandi

2. *For* Salutem dicit (*or* dat) plurimam

TO THE VERY DISTINGUISHED GENTLEMAN
MASTER GERARDUS MERCATOR OF RUPELMONDE,
RENOWNED PHILOSOPHER AND MATHEMATICIAN
and by far his dearest friend,
JOHN DEE OF LONDON
Sends hearty greetings.

It is now the eleventh year, most humane and learned
GERARDUS, from that in which I left the university, having
run through all the degrees of our schools in the seven arts
called liberal—doing so, as the proverb has it, without cork in
my nose—and undertook to travel into regions across the sea
for the purpose of investigating those sources from which, in
our age, many channels of the best of those arts have been led
to us, and of living on familiar terms with men whose lightest
single day of writing would have furnished matter enough to
require the labor of a full year for comprehension while I
formerly sat at home. And because, by the high favor of God,
I chanced at the very beginning of my travels to fall first upon
you, then busy in Louvain, and from your discussions with
me my whole system of philosophizing in the foreign manner
laid down its first and deepest roots, I now think it just and
most suitable to reason that you should claim for yourself,
as owed to your deserts, the first fruits of my labors while
abroad. And this most of all because it was the custom of our
mutual friendship and intimacy that, during three whole
years, neither of us willingly lacked the other's presence for
as much as three whole days; and such was the eagerness of
both for learning and philosophizing that, after we had come
together, we scarcely left off the investigation of difficult and
useful problems for three minutes of an hour. For the sake
of such a sincere friendship and such sweetly protracted
cooperation in philosophizing, ought we not to commend to
the eternal memory of men some "composition" or monument, so that from it a later age of scholars may be aroused

rationis, gratia, aliquod saltem σύνταγμα, vel monumentum, sempiternae hominum memoriae commendare debuimus: ut inde suavissimum illud amicitiae vinculum, quo nostri in perpetuum copulantur animi, suis quoque nectere disputationibus, postera studiosorum excitetur aetas? Et non alter alterius vel contemnere studia, vel eruditioni invidere: capita sed simul conferre, ad veri inquisitionem, & utilissimas amplificandum disciplinas. Atque ut hanc potissimum materiam, hoc tempore mihi tractandam, eligerem: penultimae tuae ad me literae, in quibus, de nobili illa, inter nos olim agitata, controversia, memoriam mihi velle refricare, videbaris, occasionem dedere. Nec in istius enodatione, seu potius demonstratione, longiorem me nunc esse, vel valetudo, quae iam per integrum annum periculosissime labefactata fuit (etiam si voluissem maxime) toleravit: vel ipsa, de Caelestium corporum virtute, Disciplina, desiderare videtur. Ex his enim quae in medium attulimus, tum ad infinitos particulares, in Arte casus, Apodictice procedendi haberi facultas potest: tum ipsa praeterea disciplinae praecipua, in his sunt iacta, confirmataque fundamenta: unde de aliis eius Artis quid sit statuendum praeceptis, industrio facile constabit artifici. Non tamen infinitas multorum & ἀναιτιολογήτας probo nugas, vel futilia decreta: quae nec ipsi talium scriptores, rationum stabilire momentis possint, nec ullus unquam alius, a Naturae viribus talia proficisci, osbervando intelligere. Tu ergo qui NATURAE observantissimus esse Cultor soles: NATURAE, in istis Aphorismis, scrutare virtutes veras, virtutes magnas, virtutes paucis vix credibiles Sapientibus, at paucissimis notas. Et ne τῶν ἀμνη[σ]τῶν τὶς,[3] suo sibi malo, ea hinc expiscari, elicereve contendat, quae illi non sunt scripta, tu cum RECEPERIS, edicas publice. Atque haec hactenus. Cum autem in literis tuis ad me, fere omnibus, quid ipse prae manibus habeam, a me scire, soles contendere: & in illis certe, quas ante nominavi, penultimis mecum egisti maxime, ut magnum illud opus

by its own disputations to form that most sweet bond of friendship by which our spirits are perpetually joined? and neither of them to despise the other's studies or to envy his learning, but to put their heads together for the investigation of truth and the expanding of useful sciences. Your next to last letter, in which you seemed to wish to refresh my memory of that noble debate formerly carried on between us, has given me an occasion to choose, in preference to all others, that subject which I am now to treat. My health, which has been dangerously shaken for a whole year now, has not permitted me, even though I wished it, to write a longer explication—or rather demonstration; nor has the subject itself, the power of the heavenly bodies, seemed to require it. For from what I have made available not only can a means be found of proceeding demonstratively in the art with regard to an infinite number of particular situations, but also, besides, the main principles of the science have been laid down and established here; so that what is to be determined about other maxims of the art will easily become clear to the assiduous practitioner. Neither do I test the infinite and "unanalyzable" trifles of many, or the worthless doctrines which those who write about them cannot themselves establish by rational processes; nor can anybody else start from the forces of nature to understand such things by observation. Do you, therefore, who are by custom a most observant investigator of nature, search out in these aphorisms the true virtues of nature: virtues which are great, and barely credible to a few wise men, but known only to a very few. And, when you receive them, I request that you declare publicly that no "incautious person" should strive to fish out and draw forth from them, to his own harm, things that are not written for him.

But enough about these matters. Since, however, you have sought in nearly all the letters which I have at hand, and have pressed me most urgently in the next to last letter I have mentioned, to publish my great demonstrative work—as you

meum Apodicticum, de Arte nova (ut tu vocas) quam primum vel in lucem darem, vel eius te ut participem facerem: me Scias, praeter periculosissimum, quo toto iam proxime elapso anno laboravi, morbum, alia etiam multa (ab illis, qui. &c.) esse perpessum incommoda, quae mea studia plurimum retardavere: viresque etiam meas, nondum posse tantum sustinere studii laborisque onus, quantum illud, Herculeum pene (ut perficiatur) requiret opus. Unde si mea haud queat opera vel absolvi, vel emitti, dum ipse sim superstes, Viro illud legavi eruditissimo, gravissimoque, qui Artium Mathematicarum unicum nobis est relictum & decus & columen: nimirum D. D. Petro Nonio Salaciensi: Illumque obnixe nuper oravi, ut, si quando posthumum, ad illum deferetur hoc meum opus, benigne humaniterque sibi adoptet, modisque omnibus, tanquam suo, utatur: absolvere denique, limare, ac ad publicam Philosophantium utilitatem perpolire, ita dignetur, ac si suum esset maxime. Et non dubito, quin ipse (si per vitam valetudinemque illi erit integrum) voti me faciet compotem: cum & me tam amet fideliter, & in artes, Christianae Reip. summe necessarias, gnaviter incumbere, sit illi a natura insitum: voluntate, industria, usuque confirmatum. Tuis igitur votis, de laborum meorum evulgandis monumentis, nondum me posse satisfacere, licet iam clare satis docui, Si tuae tamen petitioni de scriptorum meorum habendo[4] Catalogo, non responderem, merito me maximae damnares ingratitudinis. En tibi ergo eorum Titulos, quae per medias meas, maximasque difficultates, ita a me mihi composita, scriptaque extant, ut eadem (cum viribus valeam corporis, dulcique fruar otio) in publicum producere (non mihi tantum esse cognita) exoptem maxime.

4. *1558* habenda

call it—on the new art as quickly as possible, or to make you a sharer in it, you should know that, besides the extremely dangerous illness from which I have suffered during the whole year just past, I have also borne many other inconveniences (from those who, etc.) which have very much hindered my studies, and that my strength has not yet been able to sustain the weight of such exertion and labor as the almost Herculean task will require for its completion. And if my work cannot be finished or published while I remain alive, I have bequeathed it to that most learned and grave man who is the sole relic and ornament and prop of the mathematical arts among us, D. D.[1] Pedro Nuñes, of Salácia,[2] and not long since prayed him strenuously that, if this work of mine should be brought to him after my death, he would kindly and humanely take it under his protection and use it in every way as if it were his own: that he would deign to complete it, finally, correct it, and polish it for the public use of philosophers as if it were entirely his. And I do not doubt that he will himself be a party to my wish if his life and health remain unimpaired, since he loves me faithfully and it is inborn in him by nature, and reinforced by will, industry, and habit, to cultivate diligently the arts most necessary to a Christian state.

Thus I have explained clearly why I have not yet been able to respond to your wish that I should divulge the monuments of my labors. If, however, I did not respond to your request for a catalogue of my writings, you might justly accuse me of serious ingratitude. Here are the titles of the works which I have composed for myself according to my means and despite the greatest difficulties, and which stand written in such a way that I most wish them to be issued to the public (and not merely acknowledged as mine) when I am strong enough in bodily vigor and enjoy sweet leisure.

1. Perhaps for *dominus dominorum*, "master of masters."
2. Now called Alacér do Sal (in Setúba).

1. Περὶ Ἀκριβολογίας τῆς Μαθηματικῆς. opus mathematice demonstratum. lib. 16.
2. De Planetarum, Inerrantium stellarum, Nubiumque a centro terrae distantiis: & stellarum omnium veris inveniendis magnitudinibus. lib. 2. demonst.
3. De Speculis comburentibus. lib. 5. demonst.
4. De perspectiva illa qua peritissimi illustrissimique utuntur pictores. lib. 2. demonst.
5. De tertia & praecipua Perspectivae parte, quae de radiorum fractione tractat. lib. 3. demonst.
6. De Caelestis Globi amplissimis commoditatibus. lib. 2.
7. Speculum unitatis: sive Apologia pro Fratre Rogerio Bachone Anglo, in qua docetur, nihil illum per Daemoniorum auxilia fecisse, sed Philosophum fuisse maximum: naturaliterque, & modis homini Christiano licitis, maximas fecisse res: quas, indoctum solet vulgus in Daemoniorum referre facinora. lib. 1.
8. De nova Navigationum ratione. lib. 2.
9. De Anuli Astronomici multiplici usu, capita centum. liber unus.
10. De Itinere subterraneo. liber unus.[5]
11. De Trigono Circinoque Analogico. lib. 3.[6]

Aliorum adhuc tacebo nomina: qui tamen ante istorum quosdam (annuente Deo) publica frui luce possint. Hoc autem opusculum, (numero duodecimum) levi munitum armatura, tanquam Exploratorem, in varias emitto regiones: ut vera mihi doctorum proborumque hominum referat iudicia, votaque, haec a me tanta tractari, lucique promitti argumenta. Ut ex istius Exploratoris relatione, mecum & doctis cum amicis, rationem ineam, num istas meas (qualescunque) copias, in peregrinos actutum producere campos, vel domi, adhuc diligentius, in militari educare disciplina, debeam. Iam

5. *For* 10. *1558 has* De Religione Christiana. lib. 6. demonst.
6. *For* 11. *1558 has* περὶ Ἀναβιβασμῶν θεολογικῶν [Concerning Theological Fundamentals]. lib. 1.

1. "Concerning Precision in Mathematics": a work of mathematical demonstration in sixteen books.
2. Concerning the Distances of Planets, Fixed Stars, and Clouds from the Center of the Earth, and Concerning the Discovery of the True Magnitudes of all the Stars: a demonstration in two books.
3. Of Burning Glasses: a demonstration in five books.
4. Of the Perspective Used by the Most Skilled and Famous Painters: a demonstration in two books.
5. Of the Third and Chief Part of Perspective, which Treats the Refraction of Rays: a demonstration in three books.
6. Of the Great Conveniences of the Celestial Globe: two books.
7. The Mirror of Unity, or Apology for the English Friar Roger Bacon; in which it is taught that he did nothing by the aid of demons but was a great philosopher and accomplished naturally and by ways permitted to a Christian man the great works which the unlearned crowd usually ascribes to the acts of demons: one book.
8. Concerning a New System of Navigation: two books.
9. Concerning Various Uses of the Astronomical Ring: one hundred chapters, one book.
10. Concerning a Subterranean Passage: one book.
11. Concerning the Triangle and the Analogical Compass: three books.

I suppress, for the time being, the names of others, which nonetheless may, with God's consent, enjoy public light before some of these. The present little work, however—Number Twelve—furnished with light armor, I send forth into various quarters as a scout to bring back to me the true judgments of learned and honest men and their wishes that I should treat such important matters and bring them to light. The purpose is that from the report of this scout I may consider, in myself and with my learned friends, whether I should now extend my forces, such as they are, immediately

restat ut te maxime orem, egregia tua Inventa, tam in excellentissima illa Philosophiae parte, quae Physica vocatur, quam in geometricis, & geographicis rebus, publicis (quam primum queas) ut committas hominum studiis: sic enim Rempub. literariam (de qua annos ante multos, multis magnisque tuis laboribus, es optime meritus) istis utilissimis tuis, novisque Inventis, eximie profecto amplificabis. Valeas: Coeptisque tuis pulcherrimis, Deus. Opt. Max. exitus largiatur felicissimos. Iterum Vale.

Londini, anno[7] a nostro nato Redemptore
1558, Iulii. 20.

7. *1558 omits*

into foreign fields or should train them still more diligently in military discipline at home. It remains now for me to beg you earnestly to entrust to the public studies of men as quickly as you can your own remarkable discoveries both in that excellent branch of philosophy which is called physics and also in geometry and geography; for thus, certainly, you will greatly enlarge with your most useful and fresh discoveries the literary commonwealth of which, for many years, you have deserved so well by reason of your many and great labors. Farewell; and may the good and great God grant happy issues to your most beautiful undertakings. Once again, farewell.

London: July 20, in the year after our
Redeemer's birth 1558

Lectori[8]
Philosophiae sincerioris studioso,
IOANNES DEE LONDINENSIS
S. D. P.

APHORISMOS EN TIBI NOSTROS, secunda iam emitti-
mus consultatione: Numero eorundem, Ordine, vel Materia,
haud mutatis quicquam. Aphorismos, eosdem ego quidem,
Provectioribus esse scio: At in multarum magnarumque
Scientiarum cognitione non adeo progressis, longiusculos
profecto, difficilesque libros. Ex Communi, tritave philosophandi via, qui huc (Miser) diverterit, Labyrintheum esse
diversorium, actutum exclamabit. Quodcunque enim egre-
gium, in Antiquorum vel Verorum quorumcunque philosophorum experientia Theoriave fuisse aliquando positum, vel
legendo, vel meditando, vel periclitando, vel peregrinando,
Ipsemet intelligere, excogitare, invenire, audire, videreque
olim potui, Id omne, vel SELECTISSIMA QUAEQUE potius,
IN CORPUS UNUM SOLIDUM ἁρμονικῶς CONGLOBATA,
tuis hic commisi studiis. Et praeter omnium Maiorum nostrorum Inventa praeclarissima, quam Mirificis, Honorificisque
ornamentis hoc sit confertum Σύνταγμα, frequenti si tu
perquiras lectione (accuratius quaeque pensitando) certissime
conspicies. Sed tamen quae ego veritatis illustrandae, amplifi-
candaeque stimulatus desiderio (quo soli tibi essent plenissime
perspecta) nervos mei contenderim ingenioli, tu noli indignis
profanisque manifesta reddere: ne & tibi & mihi tum
dedecori, tum damno vertatur maximo. Vale amice: Mani-
busque bene precator meis.

Ex Musaeo nostro Mortlacensi,
Anno 1567. Decemb. 24.

8. *This section is new in 1568.*

TO THE READER[3]
Who is studious in the purer philosophy,
JOHN DEE OF LONDON
Sends hearty greetings.

Here are my Aphorisms, issued for you now with second thoughts, but unchanged in their number, order, and substance. I am, indeed, aware that the Aphorisms are for the more advanced; the books will certainly be rather long and difficult for those who have made less progress in the understanding of the many and great sciences. The unhappy person who turns to them from the usual and worn way of philosophizing will at once cry out that they are a confused gallimaufry. Whatever I myself have up to now been able to understand, puzzle out, discover, hear, or see by reading, or meditating, or testing, or traveling to be remarkable in what was ever put forward in the experiments or theories of any of the ancient or true philosophers I have here entrusted to your studies: all of it, or rather whatever parts are choicest, collected "harmoniously" into one solid frame. Also, if you will search diligently in repeated readings, by weighing everything with great care you will assuredly observe, besides the most excellent discoveries of all our ancestors, with what wonderful and honorable ornaments the "composition" is packed. Nonetheless you must not reveal openly to unworthy and profane persons what—driven by a yearning to illuminate and broaden the truth so that it might be fully apparent only to you—I have stretched the sinews of my poor wit to provide, lest, to your shame and mine, it should be turned to great harm. Friend, good-bye; pray wish my soul well.

From my library at Mortlake,
December 24, 1567

3. This page is in 1568 only.

Ioannis Dee Londinensis, de praestantioribus quibusdam NATURAE Virtutibus Προπαιδεύματα ἀφοριστικά.

Aphorismus 1.

UT DEUS, EX NIHILO, CONTRA[9] rationis & naturae leges, cuncta creavit: ita in Nihilum abire, rerum creatarum aliqua nunquam potest, nisi contra[10] rationis Naturaeque leges, per Supranaturalem Dei potentiam fiat.

2.

MIrabiles ergo rerum naturalium Metamorphoses fieri a nobis, in rei veritate possent, si artificiose Naturam ex pyronomiae Institutis urgeremus.[11] Naturam autem ego dico, Rem Creatam quamcunque.

3.

NOn solum ea Esse asserendum est, quae Actu in rerum natura sunt conspicua, notaque: Sed & illa quoque quae quasi Seminaliter, in naturae latebris,[12] Extare, Sapientes docere possunt.

4.

QUicquid Actu existit, Radios orbiculariter eiaculatur in singulas mundi partes, qui universum mundum suo modo replent. Unde omnis locus mundi radios continet omnium rerum in eo Actu existentium.

5.

TAm Substantia quam Accidens, suam a se Speciem exerunt: Sed Substantia omnis, excellentius multo quam

---

9. *Corrected by hand in copy text to* PRAETER
10. *Corrected to* praeter
11. *1558 ends first sentence* naturam urgeremus *and omits the second*
12. *Copy text adds in margin* atque INVOLUCRIS

JOHN DEE OF LONDON
An "Aphoristic Introduction" to Certain Especially Important Virtues of Nature

Aphorism I.

As God created all things from nothing against the laws of reason and nature, so anything created can never be reduced to nothing unless this is done through the supernatural power of God and against the laws of reason and nature.

II.

In actual truth, wonderful changes may be produced by us in natural things if we force nature artfully by means of the principles of pyronomia. I call Nature whatever has been created.

III.

Not only are those things to be said to exist which are plainly evident and known by their action in the natural order, but also those which, seminally present, as it were, in the hidden corners of nature, wise men can demonstrate to exist.

IIII.

Whatever exists by action emits spherically upon the various parts of the universe rays which, in their own manner, fill the whole universe. Wherefore every place in the universe contains rays of all the things that have active existence.

V.

Both substance and accident emit their own species from themselves, but every substance far more excellently than an accident. Also, among substances, what is incorporeal and

accidens. Et Substantiarum quidem, illa quae incorporea &
spiritalis est,[13] (vel quae Spiritalis facta est) in hoc munere
longe superat illam quae est corporea, ac ex fluxis coagmentata
elementis. Licet quanto res sunt nobiliores, tanto incomple
tiorem suam Speciem faciant: Species enim completa, idem
obtinebit nomen cum principali agente.

VI.

SIcut una res differt ab alia, ita & earundem radii differunt
in efficiendi virtute, & effectus conditione, dum circa eandem
omnino rem operantur.

VII.

RAdiorum quorumcunque ab una re in diversas emanan-
tium, diversi sunt effectus.

8.

QUicquid in aliud agit, simile quodam modo est, at alio
quidem modo dissimile prorsus illi est in quod agit, aut nulla
est actio.

9.

QUicquid in mundo est, ad aliud quid ordinem,[14] convenientiam, & conformitatem habet.

X.

QUaecunque res sunt sibi mutuo coordinatae, convenientes, vel conformatae, [15] una aliam tum sponte imitatur sua,
tum etiam aliquando una ad aliam localiter accurrit: unaque
aliam (quantum potest) tuetur & munit, etiamsi interea vis
sibi inferri videretur. Per harum ergo rerum naturalium
(modis variis) in mundo Separatim existentium, Unionem:
& aliarum Seminaliter tantum prius in Natura positarum,

13. *1558 omits parenthesis which follows*
14. *1558* ordinem habet et convenientiam
15. *1558* coordinatae, vel convenientes, una

spiritual, or becomes spiritual, far surpasses in this function what is corporeal and composed of unstable elements. Things may, however, make their own species less completely in the proportion in which they are more noble; for a perfect species is given the same name as its principal agent.

VI.

Just as one thing differs from another, so their rays differ in their power of affecting and in the causing of their effects so long as they act wholly upon the same object.

VII.

The effects of any rays pouring from one thing upon diverse things are different.

VIII.

Whatever acts upon something else is like it in some respect; but in another way it differs utterly from that upon which it acts, or there is no action.

IX.

Whatever is in the universe possesses order, agreement, and similar form with something else.

X.

Whatever things are of the same order, or harmonious, or of similar form sometimes imitate each other of their own accord and sometimes even move toward one another; one protects and defends the other as much as it can even if, at the moment, it appears to be drawing energy from the other. By the joining of such natural things that exist separately in the universe, in their differing fashions, and by the activating of other things placed somewhat higher, seminally, in nature,

Actuationem, miranda magis, vere, naturaliterque,[16] (nec violata in Deum fide, neque Christiana laesa religione) praestari possunt, quam quis mortalis, credere queat.

XI.

MUndus iste totus[17] est quasi lyra, ab excellentissimo quodam artifice concinnata: cuius chordae, sunt huius universitatis Species[18] singulae, quas qui dextre tangere pulsareque noverit, mirabiles ille eliciet harmonias.[19] Homo autem, per se, Mundanae isti Lyrae, omnino est Analogus.

XII.

SIcut lyra, constitutio quaedam est tonorum consonantium atque dissonantium, aptissima tamen ad suavissimam & infinita varietate mirabilem exprimendam harmoniam: Sic Mundus iste partes intra se complectitur, inter quas arctissima conspiciatur Sympathia: alias autem inter quas dissidium acre, atque Antipathia notabilis: ita tamen, ut tum illarum conspiratio mutua, tum istarum lis atque dissensio,[20] ad Totius consensionem atque Unionem admirandam egregie faciat.

XIII.

SEnsus nostri, non sunt sensibilium radiorum a rebus effluentium causae, sed testes.

XIIII.

SPecies non solum spiritales, sed etiam aliae naturales a rebus effluunt, tum per Lumen, tum sine lumine: non ad visum solum, sed ad alios interdum sensus, & praecipue in

16. *1558* naturaliter
17. *1558 omits*
18. *1558* res
19. *1558 omits following sentence*
20. *1558* dissentio

more wonderful things can be performed truly and naturally, without violence to faith in God or injury to the Christian religion, than any mortal might be able to believe.

XI.

The entire universe is like a lyre tuned by some excellent artificer, whose strings are separate species of the universal whole. Anyone who knew how to touch these dextrously and make them vibrate would draw forth marvelous harmonies. In himself, man is wholly analogous to the universal lyre.

XII.

Just as the lyre is an arrangement of harmonious and disharmonious tones, most apt for expressing a very sweet harmony which is wonderful in its infinite variety, so the universe includes within itself parts among which a most close sympathy can be observed, but also other parts among which there is harsh dissonance and a striking antipathy. The result is that the mutual concord of the former and the strife and dissension of the latter together produce a consent of the whole and a union eminently worthy of admiration.

XIII.

Our senses are not the causes of sensible rays flowing from things, but are witnesses of them.

XIIII.

Not merely spiritual species, but also other natural ones, flow from things both through light and without light, not to sight only but sometimes to other senses; and they come

Spiritu nostro imaginali, tanquam Speculo quodam coalescunt, seseque nobis ostendunt, & in nobis[21] mirabilia agunt.

15.

NUllus Motus perfectior orbiculari, Nec ulla forma[22] humanis exposita sensibus, LUCE est vel prior vel praestantior. Corporum igitur praestantissimorum & perfectissimorum, haec duo maxime propria erunt.

16.

QUicquid in mundo est, continue movetur aliqua motus Specie.

17.

PRo ratione motuum primorum, qui sunt caelestium corporum maxime proprii, caeteri inferiorum motus omnes naturales & excitantur & ordinantur. Moventur autem ipsa Caelestia aliquando sursum, aliquando deorsum: in anteriorem aliquando partem, aliquando in posteriorem, aliquando versus unum Mundi, vel Eclipticae polum, aliquando versus alterum.

18.

IN singulis quatuor, Maioris Mundi magnis Matricibus, sunt tres diversae partes: simul tamen concretae, conformataeque, & iustis suis contemporatae ponderibus: quas iam Notariace Å Ò Š, sive Ò Š Å, sive Š Ò Å appellare libet (Sic me enim Pyrologi intelligunt) Harum Trium proprietates effectusque[23] naturales tum principales tum secundarios tum etiam tertios, quam potes exactissime discas: Modumque reducendi

21. *1558* nos
22. *1558* qualitas
23. *1558* IN quatuor huius inferioris mundi Elementis, sunt quatuor distinctae qualitates, quae ab omnibus philosophis, primae vocantur. Quarum effectus etc.

together especially in our imaginal spirit as if in a mirror, show themselves to us, and enact wonders in us.

XV.

No motion is more perfect than circular, nor is any form exposed to human senses either more important or more excellent than light. Accordingly, these two will be especially characteristic of the most excellent and perfect bodies.

XVI.

Whatever is in the universe is continuously moved by some species of motion.

XVII.

By means of the first motions, those which are most proper to the celestial bodies, all other natural motions of earthly things are both produced and ordered. Even heavenly bodies, however, are moved sometimes up, sometimes down; sometimes forward, sometimes backward; sometimes toward one pole of the universe or the ecliptic, sometimes toward the other.

XVIII.

In the four separate great wombs of the larger world are three distinct parts; these are, however, at the same time condensed, structured, and regulated by their own appropriate weights and may now be called, by notariacal designation, Ȧ Ȯ Ṡ, or Ȯ Ṡ Ȧ, or Ṡ Ȯ Ȧ (for pyrologians will understand me if I speak so). Learn as exactly as you can the natural properties and effects of these three—not only the primary ones but also the secondary and tertiary—and the way of reducing the tertiary to the secondary and the secondary to

tertios ad secundos, & secundos ad primos: Itidem tibi est summopere examinandum, quibus casibus, eadem pars,[24] diversorum, immo contrariorum nonnunquam effectuum esse causa possit.

19.

SI duo, tria, vel quatuor Elementa,[25] & in quacunque quantitate commisceantur, ut de compositi illius vera natura, Complexione sive Temperamento[26] fias certior, per artem quandam, Graduationum dictam, tibi est elaborandum.

20.

EX qua elementorum proportione,[27] singulae humani Corporis partes, humores, & spiritus constent (quam prope fieri potest) Astrologo est pervidendum. In aliis etiam rebus naturalibus idem experiri, atque intelligere est summe necessarium, & valde iucundum.

21.

SEmen quodque in se[28] potentia habet generationis cuiusque integrum & constantem ordinem: eo quidem modo explicandum, quo & concipientis loci natura, & Circumfusi[29] caeli supervenientes vires, cooperando conspirant.

22.

SIcut primi motus privilegium est, ut sine eo torpeant omnes reliqui, sic primae & praecipuae Formae[30] sensibilis, (nimirum LUCIS) ea est facultas, ut sine ea caeterae formae omnes agere[31] nihil possint.

24. *1558* qualitas
25. *1558* Si quarumcunque qualitatum res
26. *1558 omits* Complexione sive Temperamento
27. *1558* EX qua elementorum, & primarum qualitatum proportione
28. *1558* SEmen in se
29. *1558* Circumfusi nobis
30. *1558* qualitatis
31. *1558* caeterae qualitates agere

the primary. Similarly, you are especially to consider by what contingencies the same part may be the cause not merely of diverse, but even sometimes of contrary, effects.

XIX.

If two, three, or four elements are mixed in any quantity, you are to take pains to become informed of the true nature of the compound, complexion, or temperament by an art called graduation.

XX.

The astrologer is to discover, to the best of his ability, by what proportions of the elements the various parts, humors, and spirits of the human body are composed. To investigate the same thing in other natural objects, and to understand it, is both very necessary and extremely pleasant.

XXI.

Every seed has within itself, potentially, the entire and unchangeable order of every generation. The explanation is that both the nature of the place where conception occurs and the forces of the overarching sky that fall upon the place work together to this end.

XXII.

As it is the prerogative of the first motion that without it all other motions should become quiescent, so it is the faculty of the first and chief sensory form, namely, light, that without it all the other forms could do nothing.

XXIII.

"Ότι αἱ διάνοιαι ἔπονται τοῖς σώμασι καὶ οὐκ εἰσίν αὗται μεθ' ἑαυτὰς ἀπαθεῖς οὖσαι τῶν τοῦ σώματος κινήσεων, quis philosophorum non decantat? quis mortalium non in seipso id fere quotidie experitur? Ut etiam & τοῖς τῆς ψυχῆς παθήμασι τὸ σῶμα σύμπασχον γίνεσθαι. Unde Medicus per corpus sanat animam atque temperat. Musicus autem per animam, corpori medetur & imperat. Qui ergo quamplurimis modis tum medici tum musici poterit supplere munus, is hominum & corpora & animos pro sua fere gubernaret voluntate. Verum hoc est a modestius philosophantibus, mysterii cuiusdam instar tractandum.

XXIIII.

ILla Deus in Magnete proposuit oculis mortalium spectanda, qualia aliis in rebus subtiliori mentis indagini, & sedulitati experiendi maiori, invenienda reliquit. Ego tibi vim eius attractivam primo, deinde expulsivam,[32] repulsivam, sive abactivam, tertio caelestis[33] certique cuiusdam situs appetitionem, Et quarto per solida corpora radios suos essentiales[34] traiiciendi potentiam, nunc solum in mentem redigo: alias alia eiusdem Philosophici[35] lapidis, quasi miracula (divino favente Numine) explicaturus.

XXV.

DUplices sunt stellarum omnium radii: alii sensibiles sive luminosi, alii, Secretiores sunt[36] Influentiae. Hi omnia quae in hoc mundo continentur, puncto quasi temporis[37] penetrant: illi ne adeo penetrent, quodam modo impediri possunt.

32. *1558 omits*
33. *1558 omits*
34. *1558 omits*
35. *1558 omits*
36. *1558 omits*
37. *1558 omits* puncto quasi temporis

XXIII.

"That thoughts obey bodies and do not belong among insensible things, existing as they do through bodily pertur-bations"—what philosopher does not harp on this, and what mortal does not know it through almost daily experience? as also that "The body is sensitive to the soul's sufferings." Wherefore the physician heals and regulates the soul through the body; but the musician amends and controls the body through the soul. Accordingly, whoever was able to fulfill, in a variety of ways, the office of both physician and musician could govern the bodies and minds of men almost according to his wish. But this, surely, is to be treated as a secret by discreet philosophers.

XXIIII.

In the magnet, God has offered to the eyes of mortals for observation qualities which in other objects he has left for discovery to the subtler research of the mind and a greater investigative industry. I remind you now, first, merely of its attractive virtue, then of its expulsive, repulsive, or counter-active virtue, thirdly of its desire for a particular orientation to the sky, and fourthly of its ability to penetrate solid bodies with its essential rays. God willing, I shall explain other prodigies of the same philosophical stone[4] at another time.

XXV.

The rays of all stars are double, some sensible or lumi-nous, others of more secret influence. The latter penetrate in an instant of time everything that is contained in the universe; the former can be prevented by some means from penetrating so far.

4. Not here, obviously, the "philosopher's stone" which was the end of the alchemist's experiments, but the magnet or lodestone itself.

XXVI.

STellae & vires caelestes, sunt instar Sigillorum, quorum characteres pro varietate materiae elementaris, varie imprimuntur. Quemadmodum & nostrorum sigillorum insculptae formae, facilius in unam materiam quam in aliam imprimuntur: elegantius in una, quam in alia: & tenacius in una quam in alia haerent: & in quibusdam ad quandam quasi perpetuitatem.[38] Hinc Gamaaeas considerabis attentius, aliaque maiora.

XXVII.

TAm solida quam diaphana cuncta, quae intra mundi ambitum existunt, penetrandi vis, caelestium radiorum maxime propria, magnam illis influendi, sive suas imprimendi vires facilitatem inesse demonstrat. Ut autem cum elegantia quadam, deinde cum tenacitate, vel ad infinitum fere[39] tempus retineatur immissa virtus, id ex materiae in quam influitur dispositione naturali[40] vel praeparatione artificiosa, tam in visibili forma quam in elementaribus qualitatibus & aliis, provenire debet.

28.

PRimum mobile est instar speculi sphaerici concavi, cuius qualemcunque soliditatem nullus stellarum radius sensibilis[41] penetrare potest: cum etiam[42] nullus esset talis penetrationis usus apud superos: sunt & aliae perplures demonstrationes.

XXIX.

QUascunque vires per sensibles radios, stellae efficiendo exercent, non solum directis, sed etiam fractis & reflexis illis

38. *1558 omits following sentence*
39. *1558 omits*
40. *1558 omits*
41. *1558 omits*
42. *1558 omits*

XXVI.

The stars and celestial powers are like seals whose characters are imprinted differently by reason of differences in the elemental matter. In the same way, the engraved forms of our seals are imprinted more easily upon one material than upon another, more elegantly in one than in another, and cling more tenaciously to one than to another, and to some almost permanently. You will therefore consider talismans[5] rather attentively, and other still greater things.

XXVII.

The power of penetrating everything either solid or transparent that exists within the limits of the universe—a power in the highest degree characteristic of celestial rays— proves that they possess a great readiness to influence everything, or to imprint their energies upon it. That this may happen with a certain elegance, and that the imparted virtue may be retained with some tenacity, or perhaps almost permanently, should, however, be sought by a natural arrangement or an artful preparation of the matter upon which the influence is impressed, as much in visible form as in elemental qualities and other properties.

XXVIII.

The primum mobile is like a concave spherical mirror whose solidity no sensible rays from the stars can penetrate because such a penetration would have no utility for heavenly beings. There are several other proofs of this.

XXIX.

Whatever powers the stars exercise by means of sensible rays to produce effects not only through direct rays but also

---

5. *Gamaaeas*: see the *OED*, art. "Gamahe, gamaieu." The Spanish form was *gamaeo*; the English is *cameo*.

radiis, tales suas vires ad effectus oportunos promovere possunt.

XXX.

MAgnitudines verae non solum terrestris globi, sed & planetarum fixarumque omnium stellarum, astrologo debent esse notae.

XXXI.

DIstantiae verae tam fixarum, quam singulorum plane-tarum a centro terrae, quocunque proposito tempore, astrologo constare debent: sicut & nubium, sive crassioris aëris, variae a terra altitudines.

XXXII.

QUibus terrae locis, quaecunque stella sive fixa sive er-ratica quocunque dato tempore perpendiculariter immineat: & quantum incidentiae directae angulum, cum omnibus aliis locis, supra quorum horizontes, eadem stella, eodem tem-poris momento elevatur, efficiat, cum primis est cognitu necessarium.

XXXIII.

SEnsibilem omnem radium, a stellae alicuius corpore ad punctum aliud quodcunque externum emanentem, ac cum eiusdem stellae convexa superficie aequales undique efficien-tem angulos, circumstat conus rectus, radiosus, sensibilisque: cuius Axis, ipse dictus radius erit: Vertex vero, punctum illud externum: Basis denique, convexae superficiei ipsius stellae ea portio luminosa quae dicto vertici est proxima, termi-naturque per circuli circumferentiam, ab illo termino lineae rectae (a dicto vertice ad stellam ductae) qui ipsam stellam contingit tantum, descriptam.

XXXIIII.

RAdiorum a basi luminosa alicuius stellae, ad aliquod externum punctum effluentium, Axis est fortissimus: &

through refracted and reflected ones they can apply to suitable uses.

XXX.

The true sizes not only of the terrestrial globe but also of the planets and all the fixed stars ought to be known to the astrologer.

XXXI.

The true distances of the fixed stars and of each of the planets from the center of the earth at any given time should be determined by the astrologer, as also the varying altitudes of clouds or the thicker air above the earth.

XXXII.

It is of the first importance to know what star, either fixed or wandering, hangs perpendicularly over what spots on the earth at a given time, and how great an angle of direct incidence it makes with all the other places over whose horizons the same star hovers at the same moment of time.

XXXIII.

A right cone, radiant and sensible, surrounds every sensible ray which emanates toward any external point from the body of any star and makes equal angles everywhere with the convex surface of the same star. The axis of the cone is the ray; the vertex is the external point. The base, finally, is that luminous portion of the convex surface of the same star which is nearest to the said vertex and is bounded by the circumference of a circle described by the end of a straight line drawn from the said vertex to the star and which barely touches the star.

XXXIIII.

Of the rays flowing from the luminous base of any star toward any external point, the axis is strongest; and of the reliquorum, quo ipsi axi fuerint propinquiores, eo erunt remotioribus, fortiores, respectu dicti puncti. De radiis ex profunditate stellicorum corporum egredientibus, alius nobis erit dicendi locus.

XXXV.

A Stellis terra minoribus, sensibiles cuncti qui exeunt radii directi ad terrenae convexitatis, quantam maximam possunt portionem, ab ipsarum convexarum superficierum (quae tales stellas ambiunt) portionibus veniunt, quae sunt dimidiis maiores. Et quo terrae propiores fuerint, eo a maioribus illis portionibus, radios suos directos terrae communicant: Nunquam tamen terrenae convexitatis dimidium suis illis sensibilibus radiis attingere queunt, sed portionem, eiusdem dimidio minorem.

XXXVI.

OMnes stellae terra maiores, plus quam dimidium terrenae convexitatis, omni tempore suis sensibilibus & directis radiis illustrant: Semper etiam a suae convexae superficiei portione dimidio minore, illos terrae impertiunt radios. Et quo terrae propinquiores fuerint, eo a minore tali portione radios illos directos ad terram demittunt.

XXXVII.

OMnes stellae terra minores, quanto terrae proprinquiores fuerint, tanto fortiores, eidem sui Luminis[43] radios infundunt: licet minorem eiusdem portionem sensibilibus illis[44] suis directisque radiis afficiant, quam quando sunt remotae magis.

XXXVIII.

OMnes stellae terra maiores, quanto terrae viciniores fuerint, tanto fortiores illi suos[45] imprimunt radios: & terrae

43. *1558* eidem suos utriusque generis
44. *1558 omits*
45. *1558* suos quoscunque

rest, by as much as they are nearer to the axis, by so much they will be stronger than the more remote ones with respect to the said point. We shall speak in another place of rays coming from the depths of stellar bodies.

XXXV.

From stars smaller than the earth all the sensible direct rays that fall upon as large an area as they can of the earth's convexity come from parts of the formers' convex surfaces (which encompass such stars); and these parts are greater than halves. In proportion as they are nearer to the earth they communicate their direct rays to the earth from larger areas; but they can never reach half of the earth's convexity with their sensible rays, but only a portion of the convexity which is less than half.

XXXVI.

All stars larger than the earth shine upon more than half the earth's convexity at every time with their sensible and direct rays. Also, they always send those rays to the earth from a portion of their convex surface which is less than half; and in the proportion in which they are nearer to the earth they send direct rays to the earth from a smaller such portion.

XXXVII.

All stars smaller than the earth are stronger in proportion as they are nearer to the earth and pour upon it the rays of their light, notwithstanding the fact that they affect a smaller portion of the earth with their sensible direct rays than when they are farther off.

XXXVIII.

All stars larger than the earth imprint their rays upon it more strongly in the degree in which they are nearer to it;

etiam maiorem portionem sensibilibus istis[46] suis, directisque radiis illuminant, quam quando longiori sunt semotae intervallo.

XXXIX.

PErpendendae tibi sunt cum summa diligentia Terrae & stellarum quarumcunque tum terra maiorum, tum terra minorum, portiones illae Superficiales, Sphaericae convexitatis, in stellis quidem luminosae, at in terra ab ipsis luminosis illuminatae, quae pro variis stellarum a terra intervallis, diversarum fiunt quantitatum. Et tam in terra quam in stellis terminantur per terminos superficiei conicae curtae, a linea recta, tum ipsam terram, tum ipsarum stellarum corpora contingente, descriptae. Atque de his portionibus egimus propositionibus. 35. 36. 37. & 38.

XL.

AD quodcunque punctum totius mundi venit alicuius stellae conus rectus, radiosus, sensibilisque, eiusdem coni basis, minor quidem semper erit, quam dimidium convexae superficiei ipsius stellae, cuius ille fuerit conus. Videant ergo astronomi, qua ratione stellarum metiantur diametros.

XLI.

QUanto eadem stella ab aliquo puncto totius mundi remotior fuerit, tanto sui radiosi coni recti sensibilisque basis, maior evadit, & quanto Propinquior, tanto minor.

XLII.

EXaminanda tibi erit quantitas huius basis conicae, in omni positu cuiusque stellae, respectu unius alicuius puncti, ubicunque illud punctum statuatur.

46. *1558 omits*

and they also illuminate a larger portion of the earth by their sensible and direct rays than when they are separated from it by a greater interval.

XXXIX.

You are to consider with the greatest diligence the surface portions, of spherical convexity, of the earth and of all the stars, whether greater or less than the earth: portions which in the stars are luminous, but in the earth illuminated by the stars, and which are made by the various distances of the stars from the earth to be of different sizes. These are bounded both on the earth and on the stars by the edges of a truncated conical surface described by a straight line tangent to the earth and the bodies of the stars. We have treated of these portions in propositions 35, 36, 37, and 38.

XL.

To whatever point of the entire universe the right, radiant, and sensible cone of any star comes, the base of the same cone will always be less than half of the convex surface of the star whose cone it is. Let the astronomers therefore consider by what computation they may measure the diameters of the stars.

XLI.

By as much as the same star is further distant from any point of the entire universe, by so much the base of its right, sensible, and radiant cone is larger; and by so much as the star is nearer, by so much is the base smaller.

XLII.

You are to investigate the size of this conical base in every position of whatever star with respect to any other single point, wherever the point may be located.

XLIII.

EIusdem stellae Coni recti luminosi longiores, sunt ipsis brevioribus, quibusdam de causis fortiores: at alias ob causas, longe debiliores: fortiores quidem eo videri possunt, tum quod eorum bases luminosae, maiores sunt, tum quia anguli ad verticem, minores fiant. Ex his duabus causis simul iunctis, haec nascitur ratio: Quod in longioribus conis, copiosiores radii, non incidentes solum sed etiam reflexi, magis uniuntur: unde maior vis circa talem verticem exercetur. Sed naturaliter & simpliciter, propinquitas agentis ad id in quod agit, breviores conos, fortiores efficit.

XLIIII.

QUantitatem illius convexae superficiei Lunaris, quae quocunque dato tempore, nobis illuminata convertitur, accurate elicias.

XLV.

HOrizontem nostrum verum, illum appellamus circulum, qui circumductu eius lineae describitur, cuius quiescens terminus in Mundi[47] centro fuerit, alter vero in summo statuatur caelo: ita ut a nostro vertice in huius circuli centrum demissa recta linea, eidem circulo perpendicularis existat. At Sensibilem nostrum Horizontem, alibi demonstravimus esse illam terrestris sphaerae convexam portionem, quae (omnibus super terrae uniformem convexitatem, remotis impedimentis) nobis est conspicua tota: terminaturque per circuli circumferentiam, ab illo termino lineae rectae (ab oculo nostro ad terrae contactum ductae) qui ipsam terram contingit, descriptam. Hancque portionem aliquando maiorem, aliquando minorem a nobis posse conspici, pro varia nostrae altitudinis ratione supra uniformem terreni globi convexitatem, ibidem docuimus. Ex hac quidem consideratione plurima

47. 1558 terrae

XLIII.

The longer luminous right cones of a star are, for certain reasons, stronger than the shorter ones; but for other reasons they are far weaker. They can be recognized as stronger, indeed, from the fact that their luminous bases are greater and again from the fact that their angles with the vertical are less. From these two causes together, the following principle arises: that in longer cones more abundant rays, not only incident but also reflected, are more concentrated; hence a greater force is exerted about such a vertex. But naturally and simply, the nearness of the agent to that upon which it acts makes the shorter cones stronger.

XLIIII.

You are to determine accurately the quantity of the convex lunar surface which, at any given time, is illuminated and turned toward us.

XLV.

We call our true horizon that circle which is described by the rotation of a line whose fixed point is at the center of the universe and the other end situated in the highest heaven, in such a way that a straight line dropped from our zenith to the center of the circle will be perpendicular to the circle. But I have shown elsewhere that our sensible horizon is that convex portion of the terrestrial sphere which—all impediments above the uniform curvature of the earth having been removed—is wholly visible to us and is bounded by the circumference of a circle drawn by the end of a straight line which is led from our eye to contact with the earth and which touches the earth itself. In the same place I have explained that this portion can be seen by us as sometimes larger and sometimes smaller, depending on the variation of our height above the uniform curve of the earthly globe. Upon this consideration many things depend, and these skilled persons

pendent, quae tum in Optica, tum in Astrologia, tum in Magia,[48] magni esse momenti, experientes percipient.[49]

πόρισμα

Quaecunque igitur duo caeli puncta ex diametro sunt opposita, unoquoque temporis momento in infinitis extant veris Horizontibus: Sed quaecunque duo caeli puncta, minus Semicirculo distiterint, in unico tantum haberi possunt Horizonte vero, eodem temporis articulo.

XLVI.

OMnes stellae maiores terra, ab aliqua sui portione radios sensibiles directos, ad nos mittere possunt, antequam earundem centra ad nostrum verum horizontem oriendo pervenerint: Atque ratione eadem, in occasu, sub ipso vero horizonte depressis earundem centris, nos tamen illuminare suis directis radiis possunt.

XLVII.

OMnes stellae, cum in horizonte vero alicuius loci terrestris fuerint, plus in recta linea, ab illo loco distant, quam cum supra illius loci horizontem sunt elevatae: sive uno eodemque die, sive quibuscunque diversis: modo eiusdem stellae, in illis variis temporibus, aequalis fuerit distantia a centro terrae. Alioqui enim Sol in Capricorni[50] principio oriens, longe nobis propinquior est, quam quorum imminet capitibus, in Cancro versans: & hoc propter suae eccentricitatis magnitudinem: quae etiam mutabilis est.

XLVIII.

SOlem infra nostrum verum horizontem existentem, accidentarii sui luminis radios ad nos ab aëre procurare, Crepusculinae eius Luces demonstrant: Tres igitur Superiores & fixarum plurimae, cum magis sub horizonte latent, quam

---

48. *1558 omits* tum in Magia
49. *1558 omits the corollary*
50. *1558 has the symbol* ♄.

perceive to be of great importance in optics, in astrology, and in magic.

"Corollary"

Whatever two points of the sky, therefore, are diametrically opposed appear, at any moment of time, on an infinite number of true horizons; but any two points of the sky which are separated by less than a semicircle can be contained only in a single true horizon at the same point of time.

XLVI.

All stars larger than the earth can send sensible direct rays to us from some portions of themselves before their centers come to our true horizon by rising. Also, for the same reason, when, in setting, their centers are depressed below the same true horizon, they can still illuminate us by their direct rays.

XLVII.

All stars, when they are on the true horizon of some earthly place, are more distant from the place in a straight line than when they are elevated above the horizon of that place, either on the same day or on whatever different ones, though the distance of the star from the center of the earth will be the same at the differing times. Yet the sun, when it rises at the beginning of Capricorn, is far nearer to us than to those over whose heads it hangs when it is in Cancer. This is because of the greatness of its eccentricity, which is also changeable.

XLVIII.

When the sun is below our true horizon, it furnishes rays of its accidental light to us through the air, as is shown by the brightness of twilight. Accordingly, the three superior planets and many of the fixed stars, when they lie further below the horizon than the sun does at the beginning of dawn or the end

ipse Sol, in Crepusculi matutini principio, vel vespertini fine, nobis, sui accidentarii Luminis virtutem, (licet per se non tam sensibilem quam Solis) communicabunt, instar quorundam suorum crepusculorum. Planetas etiam Sole inferiores hoc modo considerandos moneo. Fitque hoc (ut dixi) non per principalem aliquem radium, scilicet vel directum fractum vel reflexum) sed per Speciei Speciem, ut ὀπτικῆς & κατοπτρικῆς periti vulgariter loquuntur philosophi. Qua ratione Solaria Crepuscula inaequalia fiant, vide: & de aliorum planetarum Crepusculis (uti nos nunc illa appellamus) simili perquiras methodo.

XLIX.

QUa ratione stellae fixae & singuli planetae, tam infra horizontem, quam alibi constituti, ad nos vel alia terrae loca, radios[51] sui luminis, non ab ipso caelo solum, sed aëre, nubibus, aqua, montibus, & similibus corporibus reflectant, perscrutare: radiorumque caelestium fractiones multiplices attende in aëre,[52] nubibus, & aquis: Et infinitam Dei bonitatem Sapientiamque admirari & laudare cogeris.

L.

UT Stella quaelibet proprium habet nomen ex ipsius Dei impositione, Sic & naturam in se habet virtutemque[53] propriam, qualis in nulla alia, eadem omnino inveniri potest.

LI.

AD quodlibet totius mundi punctum, & quolibet temporis momento, ab omnibus stellis fixis & planetis fit talis radiorum concursus, qualis, ex omni parte similis, ad nullum aliud punctum, nec ullo alio tempore, naturaliter constitui potest.

51. *1558 has* radios tum luminis, tum Secretioris virtutis ab ipso caelo, Luminis autem *for* radios sui luminis
52. *1558 has* aëre & nubibus *for* aëre, nubibus, & aquis
53. *1558 omits* virtutemque

of twilight, will communicate the virtue of their accidental light to us—though less sensible than the sun's light—as if they had their own twilights. I urge that the inferior planets should also be considered in this way. As I have said, this is done not through any principal ray—I mean direct, refracted, or reflected—but through the species of a species, as philosophers skilled in "optics and catoptrics" commonly say. Observe why the sun's dusks are unequal, and study in the same way the dusks, as I now call them, of the other planets.

XLIX.

Investigate why the fixed stars and the various planets, both those below the horizon and those situated elsewhere, may reflect to us, or to other places on earth, rays of their own light not merely from the heaven itself but also from the air, clouds, water, mountains, and similar bodies. Observe, too, the many fracturings of the heavenly rays in the air, the clouds, and the water, and you will be driven to wonder at and to praise the infinite goodness and wisdom of God.

L.

As every star has its own name from the imposition of the name of its god, in the same way it has in itself a nature and a special virtue such as cannot wholly be found in any other.

LI.

There is such a conjunction of rays from all the fixed stars and planets upon every point of the whole universe at any moment of time that another conjunction which is in every way like it can exist naturally at no other point and at no other time.

LII.

ΚΑτοπτρικῆς si fueris peritus, cuiuscunque Stellae radios in quamcunque propositam materiam fortius tu multo per artem imprimere potes, quam ipsa per se Natura facit. Haec quidem Antiquorum Sapientum multo maxima naturalis Magiae pars erat: Et est Arcanum hoc, non minoris multo dignitatis, quam ipsa augustissima philosophorum ASTRO-NOMIA, INFERIOR nuncupata: cuius Insignia, in quadam inclusa MONADE, ac ex nostris Theoriis desumpta, tibi una cum isto libello mittimus.[54]

πόρισμα

HInc obscurae, debiles, & quasi Latentes rerum Virtutes, arte Catoptrica multiplicatae, sensibus fient nostris manifestis-simae. Unde non in stellarum solum, sed aliarum quoque rerum propriis examinandis viribus, quas per Sensibiles exercent radios, diligens Arcanorum Investigator, maximum sibi oblatum auxilium habet.

LIII.

SI quid vel Solis lumen per Lunam efficiat, vel quid ipsa ex se sola, nullis imbuta SOLIS radiis sensibilibus[55] praestare possit, cognoscere quis cupiat: ex plenilunio, & Lunae eclipsi totali cum mora, artificio catoptrico, elicere potest. Ut ad alia autem, eundem traducat experiendi modum, non opus est ut moneam.

LIIII.

QUo magis ad perpendicularitatem super aliquam ele-mentarem superficiem accedit axis radiosus alicuius stellae, eo fortius circa talem suae incidentiae locum, suas[56] vires illa stella imprimet:[57] directo quidem modo, propter maiorem

54. *1558 omits the corollary*
55. *1558 omits*
56. *1558* suas quascunque
57. *1558* influet

LII.

If you were skilled in "catoptrics," you would be able, by art, to imprint the rays of any star much more strongly upon any matter subjected to it than nature itself does. This, indeed, was by far the largest part of the natural philosophy of the ancient wise men. And this secret is not of much less dignity than the very august astronomy of the philosophers, called inferior, whose symbols, enclosed in a certain Monad and taken from my theories, I send to you along with this treatise.

"Corollary"

By this means obscure, weak, and, as it were, hidden virtues of things, when strengthened by the catoptric art, may become quite manifest to our senses. The industrious investigator of secrets has great help offered to him from this source in testing the peculiar powers not merely of stars but also of other things which they work upon through their sensible rays.

LIII.

If anybody should wish to learn what the sun's light can accomplish through the moon, or what the moon can do by itself when not steeped in the sun's sensible rays, he can find out by catoptrical skill from the full moon and from the period of darkness during a total eclipse of the moon.[6] It is unnecessary to point out that he may adapt the same kind of experiment to other problems.

LIIII.

The more nearly the radiant axis of any star approaches perpendicularity over any elemental surface, the more strongly the star will impress its forces upon the place exposed to it: directly, to be sure, because of the greater

6. Literally, "from a total eclipse of the moon, with a pause."

agentis vicinitatem: reflexo autem,[58] quia reflexi tales radii, ad incidentes, vicinius conduplicantur. Eccentricitatis ratio, in diversis zodiaci locis, planetas propiores nobis exhibere potest, cum acutissimus prorsus erit incidentiae angulus cum nostro vero horizonte, vel alia superficie. At nos, & supra de hac re diximus: & nunc significamus, in aequalibus a centro terrae distantiis, generalem hunc nos enuntiare aphorismum: esse tamen tum utilissimum, tum iucundissimum considerare exceptionis huius rationem, in variis eccentricorum circulorum locis.

LV.

QUo stellae eiusdem Mora, supra horizontem maior fuerit, eo ad suae[59] virtutis fortiorem faciendam impressionem, per directos suos radios, est accommodatior.[60]

LVI.

EX horum tantum trium diversa contemperatione, scilicet Vicinitatis, Anguli incidentiae, & Morae, ô quam multiplex consurgit ratio pro viribus eiusdem stellae exercendis, supra alicuius loci horizontem.

LVII.

MOmentaneus quilibet caeli status, tum effectus suos metit[61] infinitos, tum in aliorum eventuum Semina (congruis maturanda constellationibus) vires intendit ac imprimit efficaces.

LVIII.

OMnium caelestium motuum, ille velocissimus est, quem versus occasum, semper,[62] viginti quatuor aequalium horarum

58. *1558 followed by* per radios luminis fortius tunc operabitur, quia, *etc.*

59. *1558* suae cuiuscunque

60. *1558, 1568* accomodatior

61. *1558* maetit

62. *1558 omits*

nearness of the agent, but also by reflection, because such reflected rays are joined more closely with the incident ones. A computation of the eccentricity in various parts of the zodiac can show us which planets are nearer to us, since the angle of incidence with our true horizon or some other surface will certainly then be very acute. But I have spoken also of these matters above. I mean now to state a general aphorism with respect to equal distances from the center of the earth: that it is both useful and very pleasant to consider the reason for this exception in various places within eccentric circles.

LV.

By so much as the passage of a star above the horizon takes longer, by so much it is better fitted to make a stronger impression of its virtue by means of its direct rays.

LVI.

Because of the differing combinations merely of these three things—nearness, the angle of incidence, and pause above the horizon—O what a complex reckoning is generated of a single star's exercising of its powers above the horizon of any spot!

LVII.

Any momentary state of the heavens now reaps its infinite effects, now bends and imprints its effective strength upon the seeds of other happenings: seeds to be ripened by suitable constellations.

LVIII.

Of all heavenly motions, that is the swiftest which the periphery of the equator makes constantly toward the west spatio, aequatoris conficit peripheria: atque hunc, Diurnum Totius motum vulgariter vocant.

LIX.

QUo aequatori sunt propiores paralleli circuli, eo citatiore motu versus occasum, illorum circumferentiae, aequatoris sequuntur motum.

LX.

QUam inter se rationem habuerint, circulorum duorum quorumcunque aequatori parallelorum, circumferentiae, eandem rationem habebunt, & earundem velocitates, in diurno Totius motu: Hoc tu ad planetas & stellas fixas transfer, diurnorum arcuum respectu, &c.[63] Circumferentiae autem eam inter se habent rationem, quam ipsorum Diametri Circulorum.

LXI.

PEriodos quascunque videmus NATURAE praepotentis inviolabili lege, a caelestibus ipsis absolvi corporibus, maxima cum diligentia, a nobis animadvertendas asserimus: PERIODUM hoc loco vocamus, planetae, stellae fixae, vel alicuius caelestis puncti, ad priorem locum vel priori valde similem, per circularem motum, completam restitutionem.[64] Tempusque quod interea fluit, huiuscemodi Conversionis, Periodum nominamus.

LXII.

A Natura omnes hos illustriores[65] recipimus circulos: Horizontem, Meridianum, Aequatorem, & illi parallelos omnes: Eclipticam: Eccentricos planetarum:[66] Epicyclos, &

63. *1558 omits the following sentence*
64. *1558 omits the following sentence*
65. *1558 omits*
66. *The period which replaces the colon in 1558 appears to be a broken type*

in the space of twenty-four equal hours; and the common name for this is the diurnal motion of the whole.

LIX.

The closer the parallel circles are to the equator, the faster is the motion by which their circumferences follow the motion of the equator toward the west.

LX.

The circumferences of any two circles parallel to the equator will have the same ratio to each other that their velocities will have within the diurnal motion of the whole. Apply this to the planets and the fixed stars with respect to their diurnal arcs, etc. The circumferences have, however, the same relationships among themselves as the diameters of their circles.

LXI.

I affirm that we are to mark with the greatest diligence whatever periods we see to be completed by the celestial bodies in accordance with the inviolable law of most powerful nature. I call a period here the complete return of a planet, a fixed star, or any celestial point, by a circular motion, to its former place or a place very like it. The time that elapses in the interval we call the period of such a revolution.

LXII.

We receive from nature these more noteworthy circles: the horizon, the meridian, the equator and all the circles parallel to it, the ecliptic, the eccentric circles of planets, the alios, quos ex Theoricis planetarum, Astronomicisque Canonibus,[67] accurate discendos, monemus.

LXIII.

CIrculi omnes, Positionum (vulgariter sic[68] dicti) sunt circuli naturaliter definiti: Cum omnes illi quorundam aliorum locorum sint horizontes veri: etiamsi infiniti tales, inter horizontem tuum & meridianum statuerentur. At quo propius versus mundi polos accedis, Naturam vides quasi pedetentim istos recusare: duasque tantum ex tribus illis generalissimis, Caelestia Themata describendi viis, sibi sub polis assumere: ut & sub aequatore duas praecipue admittit: in locis autem intermediis, tres: per meridianos scilicet: circulos, eclipticae longitudinem ad rectos secantes angulos: & per istos horizontales: licet infinitis aliis modis, Natura, suarum distinguat virium proprietates.

LXIIII.

PEriodus aequatoris, est alicuius in aequatore, vel alterius puncti caelestis, ad eundem meridianum, restitutio: viginti quatuor aequalium horarum spatio, per motum Totius diurnum, absoluta. Haec autem omnium caelestium periodorum, est simplicissima, sibique semper aequalis.

LXV.

DIes naturalis, sive periodus Solis diurna, est tempus quod fluit, dum per Totius motum diurnum, Solis centrum ad eundum reducitur meridianum: Ista quidem periodus, valde inaequalis existit longitudinis.

LXVI.

ANnus tropicus solaris, est tempus periodicum, quo Sol, per proprium suum motum, ad eundem eclipticae summae[69]

67. *1558 omits* Astronomicisque Canonibus
68. *1558 omits*
69. *1558 omits*

epicycles, and others which I urge are to be learned carefully from the theorems about the planets and from astronomical canons.

LXIII.

All circles commonly said to be of positions are circles naturally defined. Since all these are the true horizons of certain other places, they will be positioned, even though innumerable, between your horizon and the meridian. But in the degree that you approach nearer to the celestial poles you see nature reject them progressively and, under the poles themselves, admit only two of the three most general ways of drawing horoscopes, as she also—especially—allows only two on the equator. In intermediate positions, however, she permits three, namely, by meridians, by circles which cut the longitude of the ecliptic at right angles, and by horizontal circles. And yet nature divides the properties of her forces in an infinite number of other ways.

LXIIII.

An equatorial period is the return of some point of the equator, or of some other celestial point, to the same meridian, and is completed in the space of twenty-four equal hours through the diurnal motion of the whole. This is the simplest of all the celestial periods, however, and is always equal to itself.

LXV.

A natural day, or the diurnal period of the sun, is the time that passes while the center of the sun is brought back to the same meridian by the diurnal motion of the whole. That period, indeed, is of very unequal duration.

LXVI.

The sun's tropical year is the periodical time within which the sun, through its own motion, is restored to the same point

locum restituitur. Huius magnitudo hac nostra aetate, observata est, dierum esse 365, horarum 5, & scrupulorum primorum 55, secundorum autem, fere 20. Mutabilem etiam huius esse longitudinem, observationes excellentium Mathematicorum exactissimae, demonstrant.

LXVII.

ANnus Solaris siderius, est tempus periodicum quod labitur interea dum Sol per proprium motum suum, ad eandem stellam fixam redit: vel ad aequalem prorsus distantiam (secundum eclipticae longitudinem) ab eadem stella fixa. Cuius magnitudinem, Thebites, Chorae filius, invenit dierum naturalium[70] 365, horarum 6, scrupulorum primorum 9, secundorum autem 20. Copernicus autem aliquanto maiorem, hoc nostro seculo esse, demonstravit: per 20 circiter secunda, scilicet.

LXVIII.

LUnares periodos veras, tum ad eandem eclipticae longitudinem, tum ad Solis coniunctionem, exacta ratione per numeros examinatas, pro quocunque dato tempore, habeas. Sunt enim inaequales valde.

LXIX.

PEriodus Lunae diurna, sive dies Lunaris, est per motum Totius diurnum, lunaris centri ad eundem meridianum restitutio perfecta: singulis pene diebus, haec, suam mutat quantitatem. Similes etiam reliquorum planetarum restitutiones ad eundem meridianum, considerantes, easdem appellabimus eorundem Dies: videlicet vel Saturni, vel Iovis, vel Martis, vel Veneris, vel Mercurii. Fixarum stellarum tardissimus motus, diei unius spatio, parum exhibebit discriminis inter suam & Aequatoris diurnam periodum.

70. *1558* naturalum

of the great ecliptic. The length of this has been observed to be, in our age, 365 days, 5 hours, 55 minutes, and approximately 20 seconds. That the length of this period, too, is variable the precise observations of excellent mathematicians show.

LXVII.

The sun's sidereal year is the periodical time that elapses while the sun, through its own motion, returns to the same fixed star or to an exactly equal distance from the same fixed star along the longitude of the ecliptic. Thebites, the son of Chora, discovered the duration of this to be 365 natural days, 6 hours, 9 minutes, and 20 seconds; but Copernicus has shown that it is somewhat greater in our century, namely, by about 20 seconds.

LXVIII.

You should know the true lunar periods both to the same longitude of the ecliptic and to conjunction with the sun, examined numerically by exact reckoning, for any given time. They are extremely unequal.

LXIX.

The moon's diurnal period, or a lunar day, is the completed return of the moon's center to the same meridian by the diurnal motion of the whole. This changes its length almost day by day. When we consider also similar returns of the remaining planets to the same meridian, we shall call them the planets' days: that is to say, of Saturn, or Jupiter, or Mars, or Venus, or Mercury. The very slow motion of the fixed stars in the space of a single day will show little difference between their diurnal period and that of the equator.

LXX.

UT Luminarium[71] secundum eclipticam, ita reliquorum quinque planetarum omnes, quas vere & naturaliter conficiunt periodos, tibi omni tempore perpendendas commendamus, tam in eccentricis, quam in epicyclis suis, per proprios suos motus. Simplices quidem per se (quantum potes) ut distinguas, compositas item seorsim, monentes.

LXXI.

UT Lunae periodicas cum sole coitiones observamus, ita & cuiusque planetae reditum ad alium tardiori affectum motu (vero quidem & proprio) quam is est, cuius restitutionem periodicam consideramus, maxima diligentia notandum significamus.

LXXII.

UT motus ille, qui est aequatoris proprius, omnium caelestium motuum est velocissimus, ita planetarum omnium periodi diurnae, sunt omnium quas vere conficiunt, brevissimo transactae tempore.

LXXIII.

EX caelestium corporum Imitatione, quae in inferioribus regulari aliqua & ordinata fieri ratione cernitur, sincerae veritatis amantes, studioseque experientes, clarissime elicere possent, quae res, vel tota, vel in sui aliqua parte, cui planetae, fixae, vel plurium stellarum colligationi subiiciatur maxime: ita ut ille planeta, stella fixa vel plurium stellarum colligatio, huius rei vel effectus, praecipuus & quasi proprius Significator (astrologorum ut utamur phrasi) possit censeri. Istam autem Imitationem variis posse modis fieri, cuivis constare credo philosophanti. Non me ergo est vel in Motu solo, Forma[72] vel

---

71. *1558* Luminarum
72. *1558 omits*

LXX.

As of luminaries relative to the ecliptic, I commend to you for consideration in this way all the periods of the remaining five planets which they truly and naturally make at every time, both in their eccentricities and in their epicycles, through their own motions. I urge you, however, to distinguish the simple periods, as well as you can, in themselves, and the compound movements separately.

LXXI.

When we observe periodical conjunctions of the moon with the sun, in the same way as the return of any planet to another which is affected by a slower, though true and proper, motion than is that of the one whose periodical return we are considering, I warn that the observation is to be made with extreme care.

LXXII.

As the motion that is proper to the equator is the swiftest of all the celestial motions, so the diurnal periods of all the planets are completed in the least time of all the movements which they truly make.

LXXIII.

From the imitation of celestial bodies which is perceived in inferior things to be made in some regular and orderly way, those who love pure truth and experiment studiously can infer very clearly what thing, either as a whole or in some one of its parts, is most subjected to what planet, or fixed star, or collocation of many stars, so that the planet or fixed star or collocation of many stars can be judged to be the principal and, as it were, proper significator (to use the astrologers' term) of that thing or effect. That the imitation can be performed in various ways I believe is known to everyone who philosophizes. It is not to be supposed that I wish

Figura, sed in aliis etiam proprietatibus & qualitatibus, hanc observari velle putandum.[73]

Consectarium. 1.

MAgus proinde industrius, Microcosmi Analogis stellaturis, ita Signata, applicando, Harmoniam experiretur maximam. Quae enim Uni Tertio conveniunt, & inter se convenientiam habere necesse est.

Consectarium. 2.

HOrum ergo Trium, duobus quibuscunque notis, quale quaerendum est Tertium, constare potest. Horumque Trium Anatomiae, singulorum propriae, sunt in reliquis duobus: Sed modo quidem diverso, scilicet Caelesti, Terrestri, vel Microcosmico. Exempli gratia, Solem, Aurum, & Cor hominis, tibi proponimus ex Anatomiae Magicae consideranda Legibus.

LXXIIII.

IN qua significatione aliquis planeta, stella fixa, plurium stellarum commixtio, vel caeli locus, praecipue excellit, ad illum Significatorem, omnes reliqui tum planetae tum fixae, in illa quidem significatione comparari debent: ut quid vel auxilii, vel impedimenti ab illis[74] recipiat in sui muneris administratione, artificiosa eliciatur indagine.

LXXV.

QUod Fixarum mutua intervalla, ex omni temporis aeternitate nunquam sunt mutata, huius elementaris mundi illis rebus, quae & sui etiam status constantem valde retinent conditionem,[75] istas maxime praeesse demonstrat. Cum tamen, & istae, Moto quodam (scilicet tardissimo) secundum

73. *1558 omits the inferences*
74. *1558 adds* aliis
75. *For* condicionem

this to be observed either in movement alone, in form, or in shape, but also in other properties and qualities.

Inference 1.

By associating things marked in this way with analogous stellar markings in the microcosm, the assiduous mage will discover a very great harmony. For things that are of the One conform also to a Third and must have some similarity between themselves.

Inference 2.

When any two of these three have been noted, what kind of third is to be sought can, accordingly, be known. The anatomies of these three—peculiar to them separately—are in the other two, but in a different way, celestial, terrestrial, or microcosmic. For example, I suggest to you the sun, gold, and man's heart as objects to be considered by means of the laws of Anatomical Magic.

LXXIIII.

In whatever signification any planet, fixed star, or group of many stars, or some place in the sky is especially eminent, all the other planets and fixed stars should be compared with that significator in that signification; so that by skillful investigation it can be inferred what help or hindrance it may receive from them in the performance of its office.

LXXV.

That the mutual spaces among the fixed stars have never been altered in the whole eternity of time shows that the stars are very much superior even to those things in the elemental universe that strongly retain an unvarying condition in their situations. Since even these, however, are borne

Eclipticae longitudinem versus orientem, tam ferantur uni-
formiter, ac si omnes uno eodemque agerentur spiritu, hoc
quidem & maximarum nostrarum rerum, seu illarum quas
iam, e nostris, maxime constantes, suique similes maxime,
iudicamus, mutationes, vicissitudinesque fieri significat. Istarum denique per Diurnum Totius motum, Circunductio: ad
totam illam caelestem constantemque Harmoniam, ex
omnibus stellis fixis resultantem, qua sibi mutuo sunt colli-
gatae (quae etiam rerum omnium quasi Forma Prima existit)
toti elementari regno, & totam cuilibet eiusdem particulae,
per principales partim suos radios, partim per accidentarios,
abundantissimo quidem modo impertiendam, (ita ordinante
Totius beneficentissimo & Sapientissimo Opifice) est instituta.
Et hoc ni esset, Nullum, ne uno quidem die (naturaliter)
praeservaretur Individuum.

LXXVI.

UT Motus Fixarum proprius,[76] generaliter nobis demonstrat, easdem talium effectuum esse causas, qui longo temporis cursu incrementa atque alterationes suscipiant suas: Sic
pro naturae proprietate,[77] quae duabus quibuscunque vel
pluribus fixis (tam ex sensibili earum radio, quam ex virtute
specifica) inest, ipsius naturam eventus, qui a duabus, vel
pluribus stellis fixis efficitur, significaturve proprie,[78] diversam
esse, est necesse.

77.

AGens debile, ut actionis fortioris specimen edat, quam
Agens simpliciter aestimatum[79] fortius, saepe usuvenit: &
hoc, aliquando propter diversitatem Subiectorum, (in quae
agunt) in dispositione sua nativa, sive artificiosa: aliquando

76. *1558* Ut fixarum motus proprius
77. *1558* varietate
78. *1558 omits*
79. *1558 omits*

so uniformly along the longitude of the ecliptic toward the east by a certain motion—to be sure, a very slow one—as if all were driven by the same spirit, this signifies that changes and vicissitudes occur even in our most important affairs, or in those of them which we now judge the most stable and most self-consistent. The revolution of the stars, finally, by means of the diurnal motion of the whole, has been established for the sake of that total and unceasing celestial harmony which reverberates from all the fixed stars: a harmony—which is a kind of first form of everything—by which they are bound mutually to each other and to the entire elemental kingdom and which shares the whole, in the fullest possible way, with each of its particles, partly through their principal rays and partly through their accidental ones, the most beneficent and wise Maker of the whole having ordained things in this way. If this were not so, no particle would be preserved naturally even for a single day.

LXXVI.

As the proper motion of the fixed stars shows us generally, the same stars are the causes of such effects as, by a long passage of time, may undergo growth and change. Thus by reason of the natural peculiarity that exists within any two fixed stars, or among many—as much because of their sensible ray as because of a specific power—it is necessary that the nature of the event which is performed or specifically signified by two fixed stars, or by many, should be diverse.

LXXVII.

It often happens that a feeble agent gives indication of stronger action than an agent which is judged to be naturally stronger, sometimes because of a difference of the subjects upon which it acts either in their native disposition or in an artfully produced one, but sometimes for other reasons. This

autem, propter alias causas.[80] Hoc maxime norunt, qui Artis
Sanctae Limina Salutarunt. Quod rite enim Septies est
Separatum, Praeparatum est, ut Septies quoque Coniungatur:
ad celeberrimam illam philosophorum Gameaeam conficien-
dam. Hoc (Dei Nutu) שְׁבְעָתָיִם Davidicum, esse, asserere audeo: ‏יב‎
quod ita nobis Dualiter expressum est.

LXXVIII.

NOn est ergo mirum, fixarum Stellarum,[81] quasdam,
quae inter illas minimae iudicantur, annis singulis certos[82]
atque sensibiles in aëre & aliis rebus, effectus producere: Tum
quod illae (licet a nobis maxime distent) terra plus octodecim
vicibus maiores existant: tum etiam, vel quod materiae, in
quam agunt, aptissimam inveniunt dispositionem: vel quod
ab aliquo planeta, earundem corroborati radii, vivaciores
quasi, firmioresque in terram torquentur: vel ab accommoda-
tissimo aliquo τοῦ περιέχοντος ad stellarum exprimendas
vires loco, adiutae, tam exiguo temporis intervallo, suarum
repetant virium effectiones. Quid de illis ergo fixis cogitare
debemus, quarum aliae totum terrestrem globum sua mole
trigesies, aliae quinquagesies quater, aliae septuagesies, aliae
octuagesies excedunt? Sed illarum (te quaeso) quae terrae
soliditatem, centies septiesque sua complectuntur magnitudine, quantam credere debemus esse efficientiam? Ab
omnibus ergo omnium ordinum fixis, divinissima per caelum
distributis harmonia, quantam quasi divinitatem simul in
terras derivari censendum?

LXXIX.

SI ex Diei naturalis tempore, deducatur una aequatoris
periodus, residuumque tempus in aequatoris partes resol-
vatur, clarissime apparebit quanta aequatoris portio, versus

80. *1558 omits remainder of paragraph*
81. *1558 omits*
82. *1568 certas*

is best known by those who have paid their respects to the threshold of the Holy Art. What is seven times properly separated is prepared also to be seven times joined for the making of that most famous philosopher's stone.[7] I dare affirm—God willing—that this is David's Seven Times,[8] 12 which has thus been expressed for us in the dual number.

LXXVIII.

It is therefore not surprising that certain of the fixed stars which are judged to be least among them produce, in particular years, definite and sensible effects in the air and other things, now because they are more than eighteen times greater than the earth even though situated at an enormous distance from us, now either because they find an especially suitable arrangement in the matter upon which they act, or because their rays are strengthened by another planet and turned more vigorously and steadily toward the earth, or because, helped by some place in their surroundings very well fitted to force out stars' energies, they repeat the action of their forces at such small intervals of time. What, then, should we think of those fixed stars of which some surpass the whole terrestrial globe in their vastness by thirty, or fifty-four, or seventy, or eighty times? And what, I ask, ought we to believe the effectiveness is of those which contain in their hugeness one hundred and seven times the earth's solidity? How much divinity, as it were, are we to think is derived by the earth at one time from all the fixed stars of all the orders distributed through the sky in their most divine harmony?

LXXIX.

If from the time of a natural day one equatorial period is subtracted and the remaining time is resolved into portions of the equator, it will appear most clearly how great a portion

7. Properly, "cameo" or "talisman," but also a stone that bore peculiar markings of apparent import.
8. "Seven Times" is in Hebrew.

occasum, vere naturaliterque, (praeter suam integram periodum) intra unius diei naturalis spatium,[83] per ascensiones
(Rectas, nominatas) promoveatur. Atque haec est vera &
propria demonstratio, illius utilissimae ac admirabilis Astro-
logicae Praxeos, quae communiter DIRECTIO appellatur
DIURNA.

LXXX.

QUando illum Aequatoris progressum Directorium,
quolibet die naturali, secundum ascensiones Solaris loci
rectas, examinaveris, tunc una etiam totius caelestis Machinae,
alium quemcunque libet, contuere locum: cuius quanta sit
facta promotio Directoria, super vel meridianum circulum,
vel horizontalem, tali loco accommodatum, interea temporis,
dum illam principalem, in Solis loco metimur, diligenter
annotabis. Directorii autem motus quantitatem, nunc per
ascensiones vel rectas vel obliquas, definimus.

LXXXI.

EX Die Lunari, subtrahas Aequatoris periodum, &
quantum, illo modo, in die una Lunari, cuncta caelestia loca,
pro ratione suarum vel rectarum vel obliquarum ascensionum,
Directorie (ut ita dicam) protrudantur, clarum evadet.

LXXXII.

PEriodus Horizontalis Diurna, planetae, stellaeve fixae,
est tempus quod fluit, dum illorum centra per motum Totius
Diurnum, ad eundem restituuntur horizontalem circulum.

LXXXIII.

EX Horizontali Solis vel Lunae periodo, unam aequatoris
periodum subtrahe: residuum,[84] illam aequatoris portionem
monstrabit, quae (praeter unam sui integram revolutionem)
versus occasum, talis periodi spatio, Directorie promovetur.

83. *1558, 1568* spacium
84. *1558* & residuum

of the equator, toward the west, is truly and naturally moved forward beyond its entire period by what are called Right Ascensions, within the space of a single natural day. And this is a true and specific demonstration of that most useful and admirable astrological happening which is commonly called diurnal direction.

LXXX.

When you examine the directional progress of the equator in any chosen natural day according to the right ascensions of the sun's position, then observe also, along with it, any other desired place of the entire celestial machine. You will mark carefully how great a directional advance is made by this above either the meridian circle or a horizontal one suited to that place in the interval of time while we measure the basic forward movement of the sun's position. We now establish the quantity of the directional motion by either right or oblique ascensions.

LXXXI.

From a lunar day subtract the equatorial period, and it will become clear how much all the celestial positions are pushed forward directionally—as I may say—in that way in a single lunar day with respect to their right or oblique ascensions.

LXXXII.

The diurnal horizontal period of a planet or fixed star is the time that passes while their centers are restored to the same horizontal circle by the diurnal motion of the whole.

LXXXIII.

From the horizontal period of the sun or moon subtract one period of the equator. The remainder will show that portion of the equator which, beyond one full revolution of itself, is advanced directionally toward the west during the space of such a period.

LXXXIIII.

LIcet Solis & Lunae, generalissimae fuerint & claris-
simae[85] vires, in hoc Directionum artificio, Reliquorum tamen
quinque planetarum (maxime in eorum propriis signification-
ibus) & Fixarum,[86] multiplices efficientiae, simili debent
observari disciplina: tam in eorum diurnis ad meridianos
reversionibus, quam ad horizontales quoscunque circulos.
Nullis[87] autem nos, aliis quam veris, nunc[88] uti stellarum
motibus memineris. Caveant[89] ergo qui vel singulis diurnis
planetarum Directionibus vel annuis (de quibus alibi agemus)
certam, eandemque praescribunt vel graduum vel minu-
torum quantitatem.

LXXXV.

PEriodi diurnae quinque planetarum, quando retrogrado
feruntur motu, aequatoris periodo sunt minores. Unde per
istos, tum aequatorem, tum alia singula mobilia caeli loca,
versus orientem postponi est necesse. Hancque aequatoris
periodi anticipationem, Veteres, Directionem conversam
appellabant. Hanc autem tam ad Meridianos quam etiam
Horizontes referri, non est necesse pluribus docere:[90] aut ex
aequatoris periodo, retropedantium periodos diurnas quascunque, auferri debere, cum satis per se sint clara.

LXXXVI.

EX IOVIS periodis diurnis, ad aequatoris periodos comparatis, vera patet & physica demonstratio Directionis cuiusdam, ab Antiquis, PROFECTIONIS ANNUAE, nuncupatae:

85. *1558* clarissime
86. *1558 omits* & Fixarum
87. *1558* Nullibi
88. *1558 omits*
89. *1558* Valeant
90. *Followed in 1558 by* aut aequatoris periodum ex retropedantium
periodis diurnis quibuscunque, auferri debere, *etc.*

LXXXIIII.

Although the powers of the sun and moon are quite general and very clear in this system of directions, the manifold workings of the remaining five planets, especially in their proper significations, and of the fixed stars, ought nevertheless to be observed by a similar method in their daily returns both to the meridians and to the horizontal circles. Bear in mind, however, that I am now speaking of no other movements of the stars than true ones. Let those beware, then, who assign a certain quantity of degrees or minutes either to the separate diurnal directions of the planets or to their annual ones, which I shall treat elsewhere.

LXXXV.

The diurnal periods of the five planets, when their motions are retrograde, are less than an equatorial period. Hence it is necessary that both the equator and the other separate movable places of the heavens be pushed back by those periods toward the east. The ancients called this anticipation of the equatorial period reversed direction. I need not explain in detail that the anticipation is to be referred as much to the meridians as to the horizons, or that from the equatorial period ought to be subtracted any diurnal periods [of planets] moving backwards, since these matters are adequately clear in themselves.

LXXXVI.

From the diurnal periods of Jupiter, when these are compared to the equatorial periods, a true and physical demonstration appears of a certain direction called by the ancients the annual progression, by which, they report, some

In qua, caelestia nonnulla[91] loca,[92] per unum circiter Dode-catemorium, versus occasum promoveri tradunt. Verum, si vel Profectionis istius partes, ad Iovis verum diurnum motum: vel ipsam annuam Profectionem integram, ad Iovis verum motum in uno anno Solari, referre velis (ut Natura te facere urgebit) clarissime tunc cernes, nec directo semper modo ista dirigi: nec eandem esse (singulis annis) graduum multitudinem, quae vel super meridianos, vel horizontes varios, pro ratione Iovialis motus veri, integrae Profectioni annuae respondet: Denique non solum quinque vel quin-decim[93] loca ita considerari posse aut debere, sed infinita fere, tam planetarum scilicet, quam fixarum, &c.

LXXXVII.

QUomodo DIRECTUS Planetae motus, non solum ad eiusdem maiorem supra nostrum Horizontem exhibendam Moram, confert, esse perpendendum: sed qua etiam ratione, intra suam Diurnam periodum, Harmonicam illam aequatoris periodum complecti, eundem facit: & ad praecipuum suum denique munus conficiendum (secundum Eclipticae scilicet Longitudinem) multo reddit habiliorem, significamus. PLANETAS ergo, cursu DIRECTO[94] progredientes, generali-ter iudicare fortiores, fortunioque quodam affectos, non est a ratione alienum. Unde motu Latos VELOCI, certissimum est, plus tum habere fortitudinis, suasque tum felicius[95] peragere significationes. Quando cum planetarum veloci cursu, etiam concurrit eorundem ad terram propinquitas maior, ex Theoricis constare tibi potest.

91. *1558 omits*
92. *1558 adds* solum quinque
93. *1558 omits* vel quindecim
94. *1558* DIRICTO (*or a broken* E ?)
95. *1558, 1568* foelicius

celestial places are moved by about a dodecatemorion (twelfth part) toward the west. Actually, if you wish to refer either the divisions of that progression to the true diurnal movement of Jupiter or the entire annual progression to the true movement of Jupiter in one solar year, as nature will press you to do, you will then perceive quite clearly that neither are these always made in a straight line nor is the number of degrees that corresponds to the whole annual progression the same in different years, by reason of Jupiter's true annual movement, above either the meridians or the various horizons. Finally, not merely five or fifteen places can or ought to be considered in this way, but an almost infinite number of places of both planets and fixed stars.

LXXXVII.

It should be considered not only how the direct motion of a planet contributes to its making of a longer stay above our horizon, but also in what proportion[9] it makes the movement to include the harmonic period of the equator within its own diurnal period. Finally, I note that this makes [the planet] much more apt to produce its peculiar effect along the longitude of the ecliptic. Accordingly, it is not opposed to reason that planets moving in a direct course should be judged stronger and endowed with a certain benevolence.[10] Thus it is most certain that when they are borne by a rapid motion they both have greater strength and act out their significations more happily. At what time greater nearness to the earth coincides with a faster movement of the planets can be manifest to you from my theorems.

9. *Or*, for what reason
10. The form of *fortunioque* is puzzling.

LXXXVIII.

PLaneta RETROGRADUS, Naturae constans decretum quodam modo perfringere videtur: periodum suam diurnam breviori absolvendo tempore, quam ipse Aequator: cuius motus, eo quod citatissimus est, sibique semper aequalis, Temporis fit nobis norma. Secundo, cum ex generali Naturae instituto, Caelestia cuncta, in motus diurni ratione, primum sequi Mobile deberent, Retrogradus autem iste planeta, (quasi sibi commissis habenis) suo nisu, primo Mobili aliquam huius sui muneris particulam praeripere videtur. Tertio, ex Diurna sua quaque periodo, aliquam illius universalis Harmoniae particulam excludit: & post aliquot elapsos dies, notabilem Totius portionem, versus ortum repulisse videbitur: quandamque magnam aequatori Iniuriam intulisse: cum ille, versus occasum,[96] perpetuo rotari debeat. Quinto, pertinax iste planeta, munus suum proprium, praecipuumque deserere videtur. Propria enim cuiusque planetae periodus, versus ortum absolvi debet. Sexto, opportunitatem illam qua ad suas fortius exercendas vires, uti poterat, (ob moram supra nostrum horizontem maiorem) recusare iudicabitur. Nec Solem igitur, neque Lunam (omnium corporearum creaturarum praestantissimas, mundoque elementari beneficentissimas, immo[97] rerum hic omnium quasi Parentes) istis implicari retropedationibus, voluit Deus. Neque reliquos quidem: nisi ad breve quoddam tempus (si ad integras eorum periodos, illud conferas) tali uti tergiversatione, patitur. Verum nullo cum NATURAE UNIVERSALIS incommodo, hoc ab istis patratur. Non magis quam acerrimae illae infinitarum pene rerum Antipathiae, NATURAE UNIVERSALIS statum ullo modo labefactant: quin ad gratissimum potius ornatum egregie faciunt: & ad NATURAE perpetuandam incolumitatem, conducunt vel maxime. EX RETROGRADATIONE tamen, particularis aliquis effectus (quem scilicet talis

96. 1558 cum occasum ille versus
97. 1558 imo

LXXXVIII.

A retrograde planet seems somehow to violate a constant rule of nature by completing its diurnal period in a briefer time than the equator itself, whose motion, in that it is the swiftest of all and always the same, stands as a temporal norm for us. Secondly, since all the celestial bodies, by a general ordinance of nature, ought to follow the primum mobile in their manner of daily motion, a retrograde planet, as if taking the bit in its teeth, seems by its own effort to snatch away some part of its office from the primum mobile. Thirdly, it removes a small part of that universal harmony from each of its diurnal periods; and after the passage of some days it will seem to have pushed back an appreciable portion of the whole toward the east and to have done the equator a serious injury, since it ought to be swung about perpetually toward the west. Fifthly, that stubborn planet appears to abandon its proper and principal office; for the proper period of every planet ought to be completed toward the east. Sixthly, it will be judged to refuse an opportunity which it could have used to exert its powers more strongly (on account of its longer stay above our horizon). Therefore God did not wish that either the sun or the moon—the most excellent of all corporeal creatures, and the most beneficent to the elemental world, being parents, as it were, of everything which is there—should be implicated in those backward movements. Or, indeed, the other planets, unless He suffers them to make such a reversal for a time that is brief if you compare it to their entire periods. Yet in truth this is performed by them with no inconvenience to universal nature. No more than those extremely sharp antipathies of a virtually infinite number of things do these [irregularities] shake in any way the status of universal nature except by making it much more pleasingly ornamented; and they contribute greatly to the preservation of nature's safety. By retrograde movement, however, a particular effect (which such a planet had evidently taken upon itself to perform) is not, for the

planeta in se receperat perficiendum) interim non promovetur, sed quasi retroagitur: Factaque Infecta fieri videntur. At quis est, qui haec, tum[98] in Politicis, tum oeconomicis negotiis esse necessaria, summeque interdum utilia, non cernat? Satiusque esse recurrere (ut dicitur) quam male currere? Iuvat ergo interdum planeta retrogradus, licet non directo ordine, sed quasi fortuito, & ex abrupto: & in contraria fere significatione.

89.

PLanetae in maximis suis a Terra distantiis (circa sua scilicet Apogaea versantes) in rebus quarum tunc fuerint proprii Significatores, fortius magnificentiusque[99] suas exercent vires, quam in eisdem faciunt, quando Terrae, circa sua nimirum Perigaea, proximi feruntur. Contra autem, in aliis sibi subiectis rebus, vivacius efficaciusque operantur, in sua maxima ad Terram propinquitate, quam in eisdem operari possunt, quando a Terra, quam queant longissime distant. Huius Aphorismi demonstratio ex 41, 43, 73, 77,[100] & aliis prius explicatis aphorismis, maximum suum & lumen & robur habet. Ut ergo in eadem, rerum per eundem planetam significatarum, specie, distincte exacteque iudicium proferas, loca maximarum & minimarum a Terra distantiarum, pro unoquoque planeta, sint tibi prius nota.[101] Per artificium autem Catoptricum, quinque planetarum Aliquem, (idque paucorum dierum Spatio[102]) longissime a TERRA distare facies: Et denuo (ictu fere oculi) ad Perigaeum, quasi Novum, deducere possis. Quosdam me olim legisse memini, in Sole Lunaque idem fuisse expertos opus. Sed videntes, ἐφρύαξαν ἔθνη. &c.

---

98. *1558* has non esse tum, *then omits* esse *after* oeconimicis (sic) *and* non *before* cernat
99. *1558 omits*
100. *1558 has* 45.73.77.
101. *In 1558 aphorism ends here*
102. *1568* Spacio

time being, assisted but is rather reversed: actions appear to have been made inactions. But who does not perceive that such things are sometimes necessary, and occasionally extremely useful, both in political and domestic business? Is it not more satisfactory, according to the saying, to run back than to advance badly? Accordingly a retrograde planet sometimes is helpful—not directly, perhaps, but as if accidentally and disconnectedly, its signification being almost contrary.

LXXXIX.

Planets situated at their greatest distances from the earth, near their apogees, exercise their powers more strongly and splendidly in matters of which they would then be proper significators than they do in the same matters when they are borne close to the earth, near their perigees. In contrast, they act more vigorously and effectively in other matters subjected to them at their greatest nearness to the earth than they can when they are as distant as possible from the earth. The proof of this aphorism is both clearest and strongest in Aphorisms 41, 43, 73, 77, and others, which have already been explained. In order to judge sharply and exactly about each kind of thing signified by the same planet, let the positions of greatest and least distance from the earth first be known to you for each planet individually. By skill in catoptrics, however, you will cause any one of the five planets to stand off very far from the earth; and that within[11] the space of a few days. And finally, in the winking of an eye, you may be able to draw it, as it were, to a new perigee. I remember reading once that certain men had tried the same work on the sun and moon; but seeing that "The heathen raged," etc. . . .

11. Less probably, "for."

90.

QUoniam Solis non est semper aequalis potentia, nec eadem significandi ratio: singulorumque etiam planetarum sint distinctae significationes, ac aliae aliaeque eorundem fiant vires, non debet idem de uniuscuiusque planetae COMBUSTIONE, pronuntiari Iudicium. Licet autem Solis excellentissima fuerit & potentissima virtus, non tamen semper laedet, dum alium planetam COMBURERE Astrologi dicunt. Fieri quidem potest, ut ille, Combusti planetae naturam ad amplitudinem quandam & magnificentiam evehat: eiusdem omne ius, in suas vires transferens. Sed dum laedit, varia est ratio. Ex Graduationum regulis, de quibus supra, aphorismo 19, egimus, quid sit omni tempore de tali Combustione statuendum (quantum ad sensibilium radiorum operationem) simplicibus semel definitis planetarum naturis, clarissime depromi potest.

XCI.

NUllus est terrestris globi locus, quem Sol, Saturnus, Iupiter, Mars, aut stella fixa quaecunque, non illustrat suo directo sensibilique radio, spatio unius suarum diurnarum periodorum, dum sub Aequatore, secundum sua vera ferantur loca. Maximum igitur est[103] huius loci privilegium: ex quo, tantillo tempore, totus terrae orbis sensibilibus directisque horum radiis illuminari, foverique possit.

92.

DUae quaecunque stellae, in locis ANTISCIIS secundum aequales, & in eandem mundi partem declinationes positae, aequales supra eundem horizontem verum, acquirunt moras. Et in aequalibus ab eodem Meridiano distantiis, omnes suae radiosae incidentiae angulos, aequales facient. Unde per motum Totius diurnum, suis radiis, istae stellae, Terrestre

103. *1558 omits*

XC.

Since the sun's power is not always the same, or the interpretation of its signification the same, and since the significations of individual planets differ and their forces vary, the same judgment should not be made of the combustion of every planet. Granted that the sun's virtue is most eminent and powerful, nevertheless it does not always injure when astrologers say that it burns another planet. Indeed, it can happen that the sun raises the nature of the combust planet to a certain fullness and magnificence by turning all the planet's juice[12] into its own energies. But when it injures, the degree is variable. From the rules of graduation which I have treated above in Aphorism 19, what is to be decided at any time about such a combustion (how much it affects the operation of sensible rays) can be derived very clearly from the simple natures of the planets once these have been defined.

XCI.

There is no spot on the terrestrial globe upon which the sun, Saturn, Jupiter, Mars, or any fixed star does not shine with its direct and sensible ray in the space of one of its own diurnal periods until they are borne beneath the equator according to their true positions. Very great, therefore, is the privilege of [the earth's] position, in that the whole orb of the earth, in so brief a time, can be illumined and warmed by their sensible and direct rays.

XCII.

Any two stars, when positioned on opposite sides of the earth in equal declinations over the same part of the earth, stay equal times over the same true horizon. When they are at equal distances from the same meridian, all the impingements of their rays will make equal angles. Accordingly, through the diurnal motion of the whole, those stars, by their

12. Or "power."

quodcunque corpus, per mutuas vices ita involvunt, implicantque, ac si, eiusdem, illis esset similis commissa cura. Ex natura ergo ita cooperantium stellarum, & intervalli earundem schematici, sive aschematici ratione, qualis ab eisdem, in (notae constitutionis) proposito corpore, sit generaliter expectandus effectus, inveniri potest.

XCIII.

LIcet caelestis cuiusque Circuli, aequatori paralleii, pars illa, quae sub Meridiano alicuius loci extiterit, (ex omnibus illius paralleli partibus) cum eius loci horizonte vero, incidentiae angulum faciat maximum: Tamen Eclipticae illa pars solum quae ab horizonte Nonagesima fuerit, altissime semper supra horizontem elevabitur. Hanc autem nonagesimam partem rarissime in Sphaera obliqua, at in Sphaera recta semper sub Meridiano inveniri, cuivis, vel mediocriter in Astronomicis versato, notissimum esse scio. Hinc in locis, quorum Vertices inter aequatorem & Mundi polos fuerint, illa Eclipticae pars, quae sub Meridiano, quocunque proposito tempore reperitur,[104] Cor caeli, appellari coepta[105] est: Nonagesima autem pars ab ascendente loco, Domus decima.

XCIIII.

STellae omnes, ut sunt Luminis participes, ita (praeter suorum insensibilium radiorum & specificas suas vires) caloris cuiusdam sunt efficientes causae.

95.

UT SOL singula caelestia corpora, sua superat magnitudine, Sic caelestis Luminis quasi fons perennis ac immensus est: calorisque nobis sensibilis, ac vitalis, praecipuus effector.

104. *1558* reperiatur
105. *1558, 1568* caepta

rays, enwrap and embrace any terrestrial body by turns, as if a similar care of the same had been committed to them. Therefore from the nature of the stars working together in this way, and from the relationship of their schematic or aschematic intervals, such an effect can be found in a body of a known constitution which is exposed to them as is generally to be expected from them.

XCIII.

Notwithstanding that the part of any celestial circle parallel to the equator which stands beneath the meridian of some place (of all the parts of that parallel) makes the greatest angle of incidence with the true horizon of that place, yet only that part of the ecliptic which is in the ninetieth degree from the horizon will always be elevated in the highest degree above the horizon. I recognize it to be well known to anybody who is even indifferently versed in astronomy that this ninetieth part can very seldom be discovered under the meridian in an oblique sphere, but always in a right sphere. Hence, in places whose vertices are between the equator and the celestial poles, that part of the ecliptic which is found beneath the meridian at any given moment has begun to be called the Heart of the Sky; but the ninetieth part from the ascendant place is called the tenth house.

XCIIII.

As all the stars are sharers of light, so, apart from the specific powers of their insensible rays, they are efficient causes of some heat.

XCV.

As the sun surpasses other celestial bodies in size, so it is— one might say—a perpetual and immense source of heavenly light and the chief producer of sensible and vital heat for us.

XCVI.

ILlum Calorem, quem Solis radiosi Coni tota Basis (ipso tunc Sole, in sui circuli Perigaeo, & in minima Eccentricitate versante) in illud terrenae Superficiei naturale punctum, quod tum sui radiosi coni vertex fuerit, tum etiam cui Sol perpendiculariter imminet, efficiendo exercet, (Doctrinae huius nostrae illustrandae gratia) esse potentiae cuiusdam, instar Sexaginta, sive Centum graduum, ponere solemus.

XCVII.

NOn potest ergo nobis ignotum esse, quanto calore suo proprio, aliud quodcunque terreni globi punctum, cui SOL in quovis alio sui Circuli loco, perpendiculariter imminere potest, afficiet: respectu illius sui maximi caloris.

XCVIII.

ET qui Solis sibi impendentis calorem, in aliqua convenienti materia apte experiri noverit, Is, non secundum proportionem solum, sed etiam secundum rei veritatem, intelliget, quantum calorem omni alteri puncto terrestri, cui imminere potest, impertiet.

XCIX.

DAta proportione inter duos caloris gradus, quos Sol in duobus diversis sui Circuli locis, in terrena loca, illi perpendiculariter subiecta, exercet: Si, quocunque dato tempore, (lucente nobis Sole) per aliquod artificium nostrum, a nobis sensibiliter excitari potest Calor, qui uni dictorum fuerit aequalis, possibile est etiam, per artificium & industriam nostram, vel eodem momento, vel alio quocunque (Lucente Sole), talem caloris gradum sensibiliter excitari, qui illi alteri sit aequalis. Ad quantam autem distantiam, hic non est explicandi locus.

XCVI.

That heat which the entire base of the sun's radiant cone brings to bear by its action, when the sun is at the perigee of its circuit and minimally eccentric, upon that natural point of the earth's surface which is at once the vertex of the radiant cone and the point over which the sun hangs perpendicularly, we are accustomed (I remark for the sake of illustrating the principle involved) to assume has a power as large as the sixtieth, or even the hundredth, degree.

XCVII.

We cannot be unaware with how much of its proper heat the sun will affect any point at all of the earth's globe over which it can stand perpendicularly in any position of its circuit in relation to its maximal heat.

XCVIII.

Also, anyone who has learned how to test fitly, in some suitable material, the heat of the sun as it stands over him will understand not merely proportionally, but in actual fact, how much heat the sun imparts at every other terrestrial point over which it can hover.

XCIX.

The proportion being given of two degrees of heat which the sun furnishes in two different places of its circuit on terrestrial points that are perpendicularly beneath it, if, at any given time, when the sun is shining upon us, there can be produced sensibly, by some kind of artifice, a heat which is equal to that at one of the two points, it is also possible, through artifice and industry, at the same moment, or at any other when the sun is shining, for such a degree of heat to be produced sensibly that it will be equal to the other. But this is not the place to explain at how great a distance.

C.

PEr hos eosdem Canones, accuratius examina, quantum reliqui planetae, a Solis virtute calefactiva deficiant, ratione basium suarum Conicarum, respectu alicuius puncti terrestris, cui perpendiculariter imminere possint, in minimis eorundem a centro terrae, distantiis. Istorum bases & distantias, ad Solis basim & distantiam comparabis: & Calores ab istis procreatos intelliges. HOC tamen memoria tu semper teneas firma: unumquemque Planetam, ex sui proprii corporis ratione, sensibilem aliam qualitatem, generali caloris commiscere[106] virtuti. Et qualis illa fuerit, non in omnibus solum planetis, sed stellis etiam fixis, (Si 53 aphorismum experiaris) per Lunam expiscari potes: & aliis etiam viis.[107]

CI.

VArietas Lunaris caloris, in quodcunque cui perpendiculariter imminere potest punctum, per Solis etiam canones cognosci potest. Scilicet, si non solum eius, a terra distantiam, sed suae etiam illuminatae partis convexae, quae ad terram convertitur, quantitatem, (instar ipsarum conicarum basium in aliis planetis) quocunque proposito tempore examinemus. Non tam apte tamen Lunares sese (ad operandum) coadiuvare radios, in Corniculari eius figura, quam cum ad orbicularem magis accedat, Cauti & diligentis Astrologi iudicio, relinquo considerandum: ut & alia in istis Aphorismis multa.

CII.

UT LUX & MOTUS sunt caelestium corporum maxime propria, ita inter planetas, SOL, LUCE propria omnes alios superat: & LUNA, proprii MOTUS pernicitate, reliquos omnes vincit. Hi ergo duo, omnium planetarum excellentissimi, merito censentur.

106. *1558, 1568* comiscere
107. *1558 adds* SOLI DEO GLORIA

C.

By means of these same rules, examine rather accurately how far the other planets lack the sun's heating power by reason of their conic bases, with respect to any point of the earth over which they can stand perpendicularly at their smallest distances from the center of the earth. You will compare their bases and distances with the base and distance of the sun and understand the heat produced by them. Only hold this steadily in mind, that each planet, by reason of its own peculiar body, mixes some other sensible quality with the general virtue of its heat. You can find out of what kind it will be not merely in all the planets but also in the fixed stars (if you will check Aphorism 53) by means of the moon, and also in other ways.

CI.

The diversity of the moon's heat upon any point over which it can stand perpendicularly can be learned also from the rules about the sun, provided, obviously, that we determine, at any given time, not only its distance from the earth but also the size of its convex illuminated part which is turned toward the earth (like the conical bases in other planets). I leave to the consideration of the wary and industrious astrologer's judgment the fact that the lunar rays do not assist one another so fitly in their working when the moon is horned as when it approaches nearer to fullness, as in these aphorisms I leave much else to the judgment.

CII.

As light and motion are the most distinctive properties of heavenly bodies, so, among the planets, the sun surpasses all the others in its proper light, and the moon is superior to all the rest in the briskness of its proper motion. Therefore these two are rightly judged to be the most excellent of all the planets.

103.

LUNA, potentissima est humidarum rerum moderatrix: humiditatisque excitatrix & effectrix.

104.

UT Solis excellentem LUCEM, praecipuum vitalis caloris moderamen comitatur: ita cum LUNAE[108] MOTU, mira quadam analogia, coniuncta est eius vis, humiditatis effectiva & moderatrix.

105.

LUNA quo terrae propinquior, & proprio motu, quo fertur velociori, eo suum in res humidas, potentius exercet dominium.

106.

SOLEM & Lunam omnium in elementali mundo nascentium & viventium, tum procreationis tum conservationis, praecipuas (post Deum) & vere physicas esse causas, ex his fit manifestissimum. Per Calidum enim & Humidum, πάντα συγκρίνεται[109] καὶ αὔξεται, (ut philosophi nostri verbis utar). Ista enim duo solum, γονιμά sunt.

107.

ANni constitutionem generalem, ex quolibet certe[110] die, per quandam analogiam, esse demonstratam videmus. Habet enim quilibet Dies naturalis, suum, tum ver, tum Aestatem, tum Autumnum, tum Hiemem.[111] Ex solo ergo Solis calore, per se partim, partim per accidens, omnes primae produci possunt qualitates, & necessario ordine. In quibus, si principia, media, finesque statuamus, Duodenarii cuiusdam rationem

108. *1558* LUNE (*for* LUNĘ)
109. *1558* συκγρινέται
110. *1558* fere
111. *1558, 1568* Hyemem

CIII.

The moon is the most powerful governess of moist things: it is the arouser and producer of humidity.

CIIII.

As a special dominion over vital heat accompanies the sun's excellent light, so, by a wonderful analogy, an effective and governing force of moisture is joined with the moon's movement.

CV.

In the proportion in which the moon is nearer to the earth and borne by a more rapid motion, it exercises its dominion more powerfully over moist things.

CVI.

From these considerations, it is manifest that the sun and moon are, after God, the chief and truly physical causes of the procreation and preservation of all things that are born and live in the elemental universe. To use the words of our philosopher, "Everything is compounded and made to increase" by heat and moisture. For these two things alone are "procreative."

CVII.

We see that the general regulation of the year is to be shown, through a kind of analogy, certainly, by any day. For any natural day has first its spring, then its summer, then its autumn, and finally its winter. From the heat of the sun only, partly through itself and partly through accident, all the primary qualities can be produced, and in the necessary order. If we distinguish beginnings, middles, and ends in these, we perceive the plan of a certain Duodenary. And it is beautiful

cernemus. Et pulchrum est considerare, quo modo tandem sub ipsis Mundi polis, ipse Annus est nisi instar Diei unius naturalis. Aphorismum istum ad altiora traducas, & maximum Secretum habes, Tu, qui Trinitatis in unitate, mysteria tractas physica: & ad Noctis multicoloris Nigredine, Opus involvendum tuum, anhelas.

CVIII.

VIginti sex diversas habitudines, quae inter fixa sidera & Solem esse possunt, pro diverso istorum & Solis in quatuor Angulis positu, ad alios etiam planetas transfer: maxime ad Lunam. Sicque consurgent, ex omnibus planetis, cum stellis fixis, hoc modo comparatis, 182 diversae rationes considerandae. Ex magnae constructionis Ptolomaei, libro octavo, has disces ad Solem habitudines.

CIX.

COrporis imperfectio, proxima & maxime propria Mortis physicae[112] causa est, non Anima. Mortis ergo naturalis, causa quoque naturalis: Ex Naturae igitur generalibus Gubernatoribus, generaliter pendet & praesignificatur.[113] In Humano certe genere, Nemo ultimum sibi a Deo praefinitum vivendi Terminum praeterire potest: Negligentia autem, paucissimi illum attingunt: Duplices unde constat humanae vitae esse Terminos.

CX.

ANima humana, & Forma uniuscuiusque rei specifica, multo & plures & praestantiores virtutes, operationesque habet, quam vel ipsum Corpus, vel eiusdem rei Materia.

112. *1558 omits*
113. *1558 omits remainder of aphorism*

to consider, finally, in what way, under the celestial poles, the year itself is nothing other than an image of a single natural day. Apply this aphorism to higher matters, you who investigate the physical mysteries in the unity of the Trinity, and you will have the greatest secret of all; and you, too, who pant after the hiding of your work by the blackness of the many-colored night.

CVIII.

Transfer also to the other planets, and especially to the moon, the twenty-six different relations which can exist between the fixed constellations and the sun according to their various positions, and the sun's, in the four angles. There will arise from all the planets, together with the fixed stars, when they are compared in this way, 182 different reckonings to be considered. Learn these relations to the sun from the eighth book of the great composition[13] of Ptolemy.

CIX.

An imperfection of the body is the proximate and most proper cause of physical death—not the soul. Accordingly, the cause of a natural death is also natural; it usually depends on the general governors of nature and is presaged by them. In humankind, certainly, nobody can go beyond the farthest bound of living predetermined for him by God; but by negligence very few reach it. Hence it appears that there are two limits for human life.

CX.

The human soul, and the specific form of every separate thing, have far more, and more excellent, virtues and operations than either the body itself or the same thing's matter.

13. The *Almagest*.

CXI.

INsensibiles, Intelligibilesve[114] planetarum radii, ad eorum sensibiles, sunt instar Animae cuiusdam ad suum Corpus.

CXII.

SIderum quaedam, eatenus MALEFICA aliquando vocantur, quatenus[115] eorundem vires in corruptam Naturam, vel male dispositam Materiam immittuntur: (Hoc nos docente Aphorismo Septimo.) Ipsa enim Sidera, per se, nihil operantur mali.

CXIII.

OMnium rerum in mundo elementari existentium, quaecunque fuerit diversitas naturalis, ea ex duabus praecipue procedit causis: scilicet ex Materiarum diversitate, & varia stellicorum radiorum operatione.

CXIIII.

OMnis res, quantumcunque exigua, in mundo elementorum existens, totius caelestis Harmoniae est Effectus: sive Exemplum quoddam & Imago. At in quibusdam rebus, hoc clarius quam in aliis apparet.

CXV.

EX ANALOGIA corporum caelestium, tam in seipsis, varie consideratorum, quam inter se mutuo comparatorum: & illorum omologa semper, in isto Elementorum regno, (ex arte a nobis supra tradita) accurate secernendo, amplissimam tu tibi viam, ad perfectam Astrologiae sapientiam, sternes.

114. *1558 omits*
115. *From here on 1558 has* eorum impressas vires, humana VOLUNTAS, in corrupta natura, ad malum exitum abire vel patitur, vel impellit: Ipsa enim sidera per se nihil operantur mali.

CXI.

The insensible or intelligible rays of the planets are to the sensible rays as is the soul of something to its body.

CXII.

Certain of the constellations are sometimes so far called maleficent as they pour their energy upon a corrupt nature or badly disposed matter, as Aphorism 7 has taught us. But the constellations themselves do no harm.

CXIII.

Whatever natural diversity there may be in all the things existing in the elemental world comes principally from two causes, namely, the diversity of matter and the differing operation of stellar rays.

CXIIII.

Everything that exists in the world of the elements, no matter how paltry, is an effect of the total celestial harmony or a particular example and reproduction of it. But this appears more clearly in some things than in others.

CXV.

By analogy with the celestial bodies considered variously both in themselves and as compared with one another, and by constantly and precisely separating their resemblances in that kingdom of the elements, by an art that has been described above, you will make smooth a broad way to a complete knowledge of astrology.

CXVI.

QUoniam Septem planetae, 120 diversas Coniunctiones nobis exhibere possunt, (scilicet dum bini coniunguntur, 21: dum terni, 35: dum quaterni 35: dum quini 21, & dum seni, 7: & dum omnes simul copulantur, 1) verissimeque summus dictet philosophus, quod ἐν αὐταῖς κεῖται ἡ γνῶσις τῶν γινομένων ἐν τῷ κόσμῳ, τῆς γενέσεως καὶ τῆς φθορᾶς: Circa illas 120 Coniunctiones, generalissimam hanc nos proponimus Methodum. Quando solum duo ex septem, copulantur, 21 variae esse possunt coniunctiones: & in illarum singulis, quis duorum planetarum fuerit fortior, considerari debet. Ex duorum ergo coniunctione, 42 diversae oriuntur considerationes: Eademque ratione, ex trium corporali coniunctione, 210: ex quatuor, 840: ex quinque, 2520: ex sex, 5040: & ex septem, 5040 variae considerationes provenire possunt. Qui omnes considerationum modi, fiunt 13692: qui tantum ex corporalibus planetarum coniunctionibus pendent: eorum etiam viribus, generalissime tantum, & non ad certos gradus (unde innumerae fere myriades, considerationum variarum, procrearentur) suppositis esse inaequalibus.

CXVII.

PEnitius Naturae virtutes introspicientes, eorumque etiam, quae superius, clarissime variisque modis, confirmavimus, satis memores, circa unamquamque mundi rem, omnium septem planetarum radios, secretioris influentiae, aut sensibiles principales vel accidentarios, omni tempore concurrere, commiscerique, certissimum esse asserimus: perpetuamque horum omnium, in rebus mundi omnibus (effectuum certe naturalium ratione, licet non secundum ipsorum vera in Caelo loca) manere coniunctionem. Unde si inaequales semper eorum essent vires, Natura 5040 modis variis generalissimis, eorum posset dispensare[116] operationes, quantum ad virium differentias. Verum, si interdum duos,

116. *1558* eorum dispensaret

CXVI.

Since seven planets can show us 120 different conjunctions (that is to say, when two are conjunct, 21; when three, 35; when four, 35; when five, 21; and when six, 7; and when all are conjunct, 1) the greatest philosopher may tell us, most truthfully, that "in these lies the knowledge of things procreated in the universe, of their origin, and of their destruction." For dealing with these 120 conjunctions, I propose the following very general method. When only two of the seven are conjunct, there can be 21 different conjunctions; and in each of these it should be considered which of the two planets is the stronger. Thus from the conjunction of two, 42 different considerations arise; for the same reason, from the corporeal conjunction of three, 210; of four, 840; of five, 2520; of six, 5040; and of seven, 5040 different permutations can be produced. All these kinds of considerations together make 13,692; and these depend merely on corporeal conjunctions of planets, on the assumption merely that their strengths are unequal, but not in specific degrees (for then almost innumerable myriads of differing permutations would be generated).

CXVII.

Looking more deeply into the virtues of nature, and remembering at the same time those things which in various ways I have clearly proved above, I affirm it to be most certain that about every single thing in the universe the rays of all seven planets—the rays of a more secret influence, or sensible rays, either principal or accidental—converge and mingle at all times, and that there remains a perpetual conjunction of all these in everything in the universe, by reason, certainly, of their natural effects, if not of their actual positions in the heavens. Wherefore, if their powers were always unequal, nature would be able to control their operations in 5040 different but general ways—as many as the

aequalibus fortitudinis numeris affici, interdum tres, interdum quatuor, interdum quinque, interdum sex, & interdum omnes (licet rarissime) consideremus: aequalitatemque istam vel in supremo vel infimo, vel intermediis posse inveniri gradibus: varios inde modos, per methodum prius explicatam, eliciemus 20295: quibus si iungamus inaequalitatis absolutae modos 5040, consurgent modi 25335, generalissimi quidem: in quibus per Graduationum regulas, philosopho est dignissimum exerceri: utilitatem enim reportabit, & voluptatem immensam. Et quo istorum duorum Aphorismorum veritatem, rationemque Logisticam intelligas,[117] praxeos nostrae quandam formulam tibi proponemus, in multis aliis etiam rebus utilissimam. Facileque poterit industrius artifex, hanc Methodum ad infinitatem quandam extendere: & non pati in Septenario solum consistere numero.* Quoniam ratio constructionis secundae partis istius tabellae difficilior videri possit: ut studiosi in hac re aliquantulum iuventur, exemplo adhibito, eandem explicabo. Si bini planetae tantum (ex septem) aequalis statuantur esse fortitudinis: inde inter omnium fortitudines, sex generalissimae habebuntur differentiae: (ut ex quinta & sexta columna patet) At per tertiam columnam, binaria coniunctio[118] inter septem planetas, 21 variis modis, alia aliaque esse potest: Et per secundam columnam, planetarum sex inaequales fortitudines, 720 modis diversis considerari possunt. Multiplico igitur 720 per 21, & prodeunt 15120. quem numerum in ultimae columnae secundo descendente loco invenies: Eadem est operandi ratio, cum tres, quatuor, quinque, vel sex, aequali

117. *1558* intelligeres, en tibi praxeos nostrae quandam formulam, in multis aliis, *etc.*

118. *1568* coniuctio

* In the original, the chart on page 194 is placed here.

differences of their powers. Indeed, if we should consider that sometimes two are furnished with equal degrees of power, sometimes three, sometimes four, sometimes five, sometimes six, and sometimes—although very rarely—all, and that such an equality can be found in the highest degree, or in the lowest, or in intermediate degrees, we shall arrive at 20,295 varying relationships by the method already explained. If to these we should add 5040 kinds of absolute inequality, the result would be 25,335 kinds—very general ones, to be sure—in which it is most worth while for the philosopher to be trained, through the rules of graduation, since he will obtain use and immeasurable delight from them. In order that you may understand the truth and the arithmetical ground of these two aphorisms, I add a table which shows the calculations and is very useful for many other purposes. An industrious workman will be able to extend this method easily almost to infinity and not allow it to be limited to the septenary number.* Since the system underlying the construction of the second part of this table may be more difficult to see, I will offer an example and then an explanation so that concerned persons may be helped a little. If only two planets of seven are determined to be of equal strength, six general distinctions will be found among the strengths of all, as appears from the fifth and sixth columns. But according to the third column, a binary conjunction among seven planets can occur in twenty-one different ways; and according to the second column six unequal strengths of planets can be considered in 720 different ways. I therefore multiply 720 by twenty-one and get 15,120, a number that you will find in the second place down in the last column. The same is the principle of computation when three, four, five, or six are assumed to be furnished with equal

---

* In the original, the chart on page 195 is placed here.

Praxeos Formula⋆

	Pro Aphorismo CXVI				Pro Aphorismo CXVII		
I		I	0	0	0	7	5040
2	/	2	21	42	2	6	15120
3	/	6	35	210	3	5	4200
4	/	24	35	840	4	4	840
5	/	120	21	2520	5	3	120
6	/	720	7	5040	6	2	14
7	/	5040	I	5040	7	0	I
				13692			25335

| Planetae inaequalis fortitudinis coniuncti corp. | | Transpositiones secundum inaequalitatis generalia discrimina. | Varietates coniunctionum binorum, ternorum, &c. | Coniunctionum variarum per coniunctorum numerum transpositum, Multiplicationes. | Planetae aequalis fortitudinis coniuncti. | Inaequalitas, ex aequalitate, producta. | Aequalitates secundum coniunctiones variatae: & inaequalitatis (ex aequalitate profectae) transpositione, considerationum rationes variae. |

⋆ *1558 omits these two words*

Formula of Procedure

		For Aphorism XCVI			For Aphorism XCVII		
I		I	0	0	0	7	5040
2	/	2	21	42	2	6	15120
3	/	6	35	210	3	5	4200
4	/	24	35	840	4	4	840
5	/	120	21	2520	5	3	120★
6	/	720	7	5040	6	2	14
7	/	5040	I	5040	7	0	I
				13692			25335

Conjunct corporeal planets of unequal strength.

Permutations according to general distinctions of inequality.

Numbers of different conjunctions of two, three, and more planets.

Multiplication of differing conjunctions by permutations of the conjunct bodies.

Conjunct planets of equal strength.

Inequality drawn from equality.

Equalities varied according to conjunctions, and various reckonings of permutations by transposition of inequality (derived from equality).

★ Error for 126 (hence the total is also wrong).

supponantur esse praediti fortitudine: Denique ad ampliorem huius rei explicationem, en tibi brevissimam operis formulam.

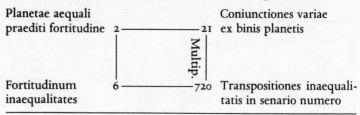

Planetae aequali
praediti fortitudine   2 —————————— 21   Coniunctiones variae ex binis planetis

Multip.

Fortitudinum   6 —————————— 720   Transpositiones inaequali-
inaequalitates   tatis in senario numero

Productum. 15120[119]

CXVIII.

CUm in anni alicuius Solaris Revolutione, alterius planetae periodi notabilioris[120] principio, vel quocunque alio tempore, fortis aliqua & rara[121] in caelo fuerit vel planetarum inter se, vel planetarum cum fixis, configuratio: vel Phaenomenum[122] inusitatum Meteorologicum, Per totum terrae orbem astronomice Circumspice, Quis terrestris[123] locus fortissimam propriamque caeli figuram, in quocunque velis significato, talis configurationis, Apparitionisve[124] primae momento, obtineat, vel obtinere possit. Hinc enim non solum[125] a stellarum, aliorumque caelestium, & Sublimium naturis, eventus illius Loci proprios maxime, sed ab eventibus egregiis Locorum terrae particularium, proprias planetarum, fixarumve[126] & aliorum caelestium, Sublimiumque eliciendi naturas, modus datur insignis, secretusque. Hinc etiam Sapiens, (modo Cosmopolites esse possit) nobilissimam Scientiam haurire potest: sive de prosperis procurandis, sive removendis noxiis: vel e contra: tam sibi quam aliis. Locorum terrestrium Opportunitas, tanti est momenti.[127]

119. *1558* 11520
120. *1558 omits*
121. *1558* notabilisque *for* et rara
122. *1558 omits* vel Phaenomenum inusitatum Meteorologicum
123. *1558 omits*
124. *1558 omits* Apparitionisve primae
125. *1558 continues* naturis, eventus proprios maxime, *etc.*
126. *1558 continues* eliciendi naturas, modus datur, *etc.*
127. *1558 omits the following Annotatio*

strength. Finally, as a fuller explanation of the matter, here is the briefest possible formulation of the operation.

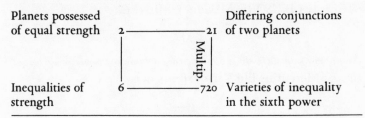

| Planets possessed of equal strength | 2 ——————— 21 | Differing conjunctions of two planets |
| Inequalities of strength | 6 ——————— 720 | Varieties of inequality in the sixth power |

Product 15120

CXVIII.

When, in the revolution of some solar year, at the beginning of a rather noteworthy period of another planet, or at any other time, there is some strong and unusual configuration in the sky either of planets only or of planets and fixed stars, or some extraordinary astronomical phenomenon, look about astronomically through the whole orb of the earth to see what terrestrial place is, or might be, affected by an extremely powerful and special figure of the sky, of any signification at all, at the moment of the configuration's first appearance. For in this way a remarkable and secret method is given of determining not only—from the natures of stars and other celestial and exalted bodies—the happenings peculiar to that place, but also, from the remarkable events of particular places on earth, the special natures of planets, fixed stars, and other celestial and exalted bodies. Thus also a wise man (if only he can be a citizen of the world) can drink in a most noble science for the purpose either of procuring good fortune or of removing bad, or contrariwise, as much for himself as for others. Of such great importance is the convenience of terrestrial places.

Annotatio.

SIc illos *Circumspexisse Magos est verisimile, qui olim dixerunt,*
STELLAM EIUS VIDIMUS IN ORIENTE.

CXIX.

Χωρὶς τῆς κοσμικῆς συμπαθείας, τοῖς ανθρώποις οὐδὲν ἐπιγίνεται:
ut nos Mercurius ille Termaximus docuit.

120.

Ἱκανὰ τὰ ΘΕῖΑ, καὶ ἡ τούτων Περιφορὰ, τὴν ἐν τῷ κόσμῳ τῶν
φυσικῶς γινομένων, Συνεχείαν φυλάσσειν.

SOLI DEO HONOR ET GLORIA

Comment

It is probable that those Mages looked about thus who long ago said, *We have seen His star in the east.*

CXIX.

"Nothing happens to men without cosmic sympathy"
—as Thrice-Great Hermes has taught us.

CXX.

"Suitable divine things and their revolution are sufficient to preserve the continuity of everything generated physically in the cosmos."

HONOR AND GLORY TO GOD ALONE

SVPERCAELESTES
RORETIS AQVAE.

ET TERRA FRVCTVM
DABIT SVVM.

QVATER △ NARIVS
IN TERNARIO CONQVIESCENS.

Excuſum Londini apud Re-
gtnal·dum Vuolfium, Regiæ Maieſt.
in Latinis Typographum.

ANNO DOMINI M.D.LXVIII.
Ianuary. 9.

LET THE
WATERS ABOVE THE   AND THE EARTH WILL
HEAVENS FALL   YIELD ITS FRUIT

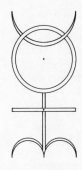

THE QUATERNARY
RESTING IN THE TERNARY

Printed at London
By Reginald Wolfe, The King's Majesty's
Latin Printer
A.D. 1568
January 9

✎ General Notes

The notes have two main purposes: first, to assist the general reader by offering nontechnical explanations of statements more fully explained in the *Introduction*, and, second, to relate Dee's astronomical ideas to the occultist thought of the period, which as Dee aged was to occupy more and more of his attention. References to the *Introduction* are included where they are appropriate. The diagrams offered in the notes are meant to be visually suggestive; those in the *Introduction* are geometrically proved.

I am grateful to John Heilbron for his help, which has corrected a number of astronomical misunderstandings and assisted a more accurate or more economical phraseology in many places. I take full responsibility for whatever errors remain.

WAYNE SHUMAKER

TITLE PAGE

The motto appears to be Scriptural but is so only indirectly. *Isaiah* 45:8 begins *Rorate, caeli, desuper* ("Drop down, ye heavens, from above"), and *Psalm* 66:7 includes *Terra dedit fructum suum* ("The earth has yielded its fruit"). But in the King James version the line appears as *Psalm* 67:6, where it is translated "Then shall the earth yield her increase." *Zachariah* 8:12 has *vinea dabit fructum suum, et terra dabit germen suum, et caeli dabunt rorem suum* ("the vine shall give her fruit, and the ground shall give her increase, and the heavens shall give their dew"). The Latinist will note that *rorem*, "dew," is cognate with *rorate*, "drop down." Dee probably intended a comparison between astrological influence and dew.

The design represents Astronomy (identified by the star) and her influence (the feather with which she writes her decrees and the swirls of "influence" which converge upon the universe). The complicated symbol enclosed by the swirls and streamers is the subject of Dee's *Monas Hieroglyphica* (Antwerp:1564) and could be fully explicated only in a separate essay. At the top is the symbol of the moon, in the center the symbol of the sun, beneath that a cross, and at the bottom the symbol of Aries, thought to have been in the ascendant at the first vernal equinox after creation. Other planetary symbols are also present, as follows: Saturn, ♄ ; Jupiter, ♃ or ♃ ; Mars,☉+ (for the more usual ♂); Venus, ♀ ; and Mercury, ♃ (evidently alternative to ☿ , which is also present). The cross is the Christian cross, but it also signifies the ternary (two straight lines and the point of intersection) and the quaternary (four right angles and four straight lines). All this is explained in the *Monas*.

On the 1558 title page, the motto at the top asserts rather peremptorily the authority of a mage who, like Dee, has studied the most reliable sources. The passage from *Luke* 21:25 is one of those frequently cited as offering Scriptural foundation for astrology. The motto in the scroll at the left hints at the significance of the central symbol, the monad. The right-hand scroll affirms the astrological powers of all the planets. Mars, as usually symbolized, is more obviously "endowed with a sting" than Mercury (unless the symbol of Aries, ♈, at the bottom, is seen as an arrow head), but the notion seems to be that all the planets shoot rays at the earth. On the architraves and plinths of the columns are the names of the four "qualities" of the planets: Mars is hot and dry, Saturn cold and dry, and so on.

"TO THE VERY DISTINGUISHED GENTLEMAN"

The primacy given here to continental scholarship over English is striking: one day's writing by a member of Mercator's group would require a year's study in England. Dee's commitment to his program is vividly described and his gratitude for Mercator's help expressed. For the rest, the letter is partly apologetic and partly boastful. Dee submits his unpublished writings to a literary executor, the mathematician Pedro Nuñes, for editing if his own life should be cut short; and at the end he acknowledges that the reception of the *Propaedeumata* will reveal to him whether his period of apprenticeship is now ended. Yet he already regards himself as an advanced thinker. The *Propaedeumata* itself is a "monument"; he has scorn for the "worthless doctrines" of many writers; he notes Mercator's expectation that some future treatise of his will be a "great demonstrative work"; and he lists with satisfaction the titles of eleven other works by himself which remain unpublished.

The astronomical ring of the ninth title was an ancient device consisting of three circles, two fixed ones representing the equator and the meridian and one pivoted to rotate about the world's axis. When the instrument is properly aligned, the right ascension and declination of any visible star can be measured. Use of the ring seems to have been revived by Gemma Frisius. A picture of one to Mercator's design may be found in H. Michel, *Scientific Instruments* (1967), Plate 47. See also *Annuli Astronomici ... Usus, ex variis authoribus*, by Gulielmus Cavellat (1558).

"TO THE READER"

The modern reader of the *Propaedeumata* may derive consolation from Dee's acknowledgment that educated contemporary readers also had trouble with it. The "usual and worn way of philosophizing" is the Aristotelian way, which remained standard at the universities. Note the emphasis on "ancient or true philosophers" and "the discoveries of all our ancestors." The prestige of ancient science, as of ancient philosophy, rhetoric, ethics, etc., remained very great. "Innovation" was a bad word, and original discoveries were often supported by citations from antiquity. The warning at the end against the revelation of hidden truths to unfit readers is commonplace and suggests that Dee's text had not merely contemplative but also operational (i.e., astrological) value.

I

The handwritten corrections of CONTRA to PRAETER ("against" to "beyond")—made, presumably, by the Nicholas Graye who owned the copy text—have, no doubt, a theological import. No great astuteness is required, however, to defend CONTRA on the ground that human reason only is meant, and that "nature" refers to the way things have been since their creation.

In a letter of August 7, 1574, to William Camden, Dee asserted that this aphorism provided a firm foundation for all the rest: "At in primo statim aphorismo meo (ut caetera taceam) artis cuiusdam magnae jactum est fundamentum firmum, quae (brevissime) iisdem meis explicata est Aphorismis" (Bodleian, MS. Ashmole 1788, f.72, quoted by Clulee, *Glas of Creation*, p. 74n.). To a modern reader, the aphorism seems roughly equivalent to a law of conservation. In the sentence just quoted, however, Dee calls it the basis of "a certain great art," a phrase often used of alchemy.

II

Pyronomia is defined by Andreas Libavius as "the science of using and regulating heat and fire in one's operations" (*Alchemia*, 1597, I, xiiii, 24–25). A following passage describes four degrees of heat, like those taken for granted by Dee's discussion of the mixture of qualities in the "Preface." As yet there were no thermometers. Although the "wonderful changes" would include alchemical transformations, Dee's main purpose is to acknowledge that despite the indestructibility of matter its forms can be changed.

III

A truism then as now, but perhaps intended as preparation for the discussion of the insensible rays mentioned in XXV.

IIII

These rays are sometimes invisible (again see XXV). Even the dimmest of them reach to the outer limits of the universe.

V

"Substance" and "accident" are technical terms of Aristotelian metaphysics. "Primary substance" signifies an individual thing—this man, that tree. "Secondary substance," in the sense used here, signifies the essence of a thing, the species to which it belongs: rational-animalness, oakness. "Accident" is a characteristic of an individual that helps to define the individuality without altering its species: baldness, fatness, being-in-motion; leafiness, ten-feet-tallness, having-many-acorns. Each individual thing or primary substance may, for purposes of Aristotelian analysis, be considered a compound of accident with secondary substance or essence. For "species" see above, pp. 61–64.

VI

The various heavenly bodies, and especially the planets, have differing astrological influences and produce them differently.

VII

Thus the sun's heat might liquefy ice but vaporize moisture. Again, radiation from Mars might hearten a timid man but enrage a choleric one.

VIII

There must be some "sympathy," as between the magnet and the object it attracts. For sympathies and antipathies more generally, see Note X. Lack of difference would result in identity and the absence of "action."

IX

The basic notions here are those of orderliness and correspondence. An example of correspondence is given in LXXIII, Inference 2, where the sun is said to be to the heavens what gold is to the earth and the heart to a man.

X

A consequence of correspondence is again "sympathy," which was important to current medicine. For the medical doctrine, see Hieronymus Fracastorius (Girolamo Fracastoro), *De Sympathia & Antipathia Rerum* (1550). In view of the mention of "things placed somewhat higher," it is possible, though far from certain, that Dee had in mind also the increasing of astrological influence by talismans of the kind Marsilio Ficino was drawn to but feared. For a discussion of these, see Wayne Shumaker, *The Occult Sciences in the Renaissance* (1972), pp. 127–130. More important, probably, is the use of sympathies in alchemical operations. In any event, "natural" forces only are in question, not daemonic. "Seminal" things are especially powerful. Such natural magic, Dee believes, is in no way unchristian because it utilizes powers inherent in God's creation.

XI

As is well known, the notion of a cosmic harmony goes back at least to Pythagoras (sixth century B.C.), who thought that the planetary intervals were such that the movements of the planets in their orbits made ravishing, though inaudible, music. In Christian thought, its inaudibility to man was caused by the Fall. See S. K. Heninger, Jr., *Touches of Sweet Harmony* (1974), especially pp. 179–189. As the microcosm, man also is a harmonious little world patterned on the cosmos. For "touching" the strings of the universal lyre, see D. P. Walker, *Spiritual and Demonic Magic from Ficino to Campanella* (1958), I, i:2, "Ficino's Astrological Music"; also Marsilio Ficino, "De vita caelitus comparanda" (Part III of *De vita triplici*), summarized in Shumaker, *Occult Sciences*, pp. 132–133. Cornelius Agrippa explains "what sounds accord with each star" (i.e., planet) in *De occulta philosophia*, II, xxvi.

XII

An example of "dissonance" and "antipathy" might be the retrograde movement of a planet, which appears to carry it for a time contrary to its dominant course. Such irregularities, however, cause "no inconvenience to universal nature" but rather make it "more pleasingly ornamented" and contribute to its stability (LXXXVIII).

XIII

The statement now appears otiose but may not have seemed so at a time when the eyes were thought to emit rays.

XIIII

For "species" see above, pp. 61–64. The phrase, "emit its species upon" is roughly equivalent to "impress its power upon."

"Imaginal spirit" is the part of the mind that registers images, i.e., the sensorium. Our senses receive images—not merely visual—through both the sensible and insensible rays mentioned in XXV and are affected by them, sometimes with surprising force. The images produced by the invisible rays are presumably not visual.

XV

The superiority of the circular or spherical form over other forms was already commonplace in antiquity. In

Christian thought, supralunary nature was perfect because the Fall affected only what lay within the moon's sphere. *Prior*, here translated "important," may mean rather "antecedent" in the order of creation.

XVI

The fondness of premodern times for stasis, which implied permanence and was sought, instead of "progress," by all Utopian schemes of social reorganization down to the time of James Harrington's *Oceana* (1656) and beyond, was compromised by a sense that immobility suggested death. A resolution was found in the recognition that perpetually recurrent changes allowed for both stasis and movement. Both were found not only in the natural day and the seasons but also in the heavens, where the movements were repeated and the precession of the equinoxes, in time, would reproduce the original celestial configurations. A more literary notion of patterned movement within stability was that of the Cosmic Dance.

XVII

The responsibility of the first motions for all the rest, which is Aristotelian, is implied by the concept of primacy. The movements mentioned in the second sentence, which complicate the astronomer's calculations, do not compromise the universal order.

XVIII

This is the most inscrutable of all the aphorisms, and no more than hints of its possible meanings can be given.

What the matrices or wombs may be is suggested by a comment of Allen G. Debus in *The English Paracelsians* (1965), p. 28: Paracelsus sometimes "spoke of the four elements on the highest level in their cosmic sense, as imperceptible elements or matrices." On the bodily level, however, "they are represented as perceptible elementary bodies in terms of the four concentric spheres of earth, air, water, and fire." The system, which is standard, is illustrated on the title page of Robert Fludd's *De supernaturali, naturali, praeternaturali et contranaturali microcosmi historia* (1619), where the realms of earth, water, air, and fire appear inside the moon's sphere. The system was Aristotelian. The arrangement, from the earth outward, is in the order of increasing lightness. The extension of the system to the supralunar regions, if that is what Dee intended, would be surprising; but the "larger world" is certainly the universe.

The "three distinct parts" within each of the matrices are uncertain. One possibility is offered by the three "principles" of Paracelsus, sulphur (the principle of combustibility, substance, and structure), salt (that of solidity and color), and mercury (that of a "vaporous quality"); cf. Debus, *English Paracelsians*, p. 27. The relevance of this schema is doubtful.

The regulation of the three parts "by their own appropriate weights" offers no problem. Much was made of a text from *Sapientia* 11:21, "omnia in mensura, et numero, et pondere disposuisti" (in the King James Bible, *Wisdom* 11:20, "thou hast ordered all things in measure and number and weight"). Conceivably the three elements designated Å Ó Ṡ (if they are elements) are composed of different combinations (by weight and degree of compression).

The next to last sentence is obscure in detail but clear in general purport. By some process of analysis the secondary and tertiary properties (or "elements") are to be found derivative from the primary, which can then be understood as containing the others *in ovo*, as the *prima materia* or First Matter contained all matter potentially within it.

The initials, with their curious dots, are impossible to interpret with confidence. Although they might be cabalistic, I have found nothing comparable in Johann Reuchlin's *De Arte Cabalistica* (1517), the most likely source. *Notariacè*, from νοταρικόν, has to do usually with abbreviations (especially initials, as in SPQR for "Senatus Populusque Romanus") or with the numerological values of letters. The anonymous *Sefer Yeẓirah* or *Book of Creation*, the earliest extant Cabalistic writing, uses the Hebrew equivalents of A, M, and S to represent air, water, and fire, respectively (III:5). (I owe this reference to David Winston of the Graduate Theological Union, Berkeley.) Conceivably Dee's Ȧ and Ṡ might stand for air and fire and his Ȯ for water or earth.

It does not help to search for Latin words beginning with A, O, and S. For example, A might represent *aurum* (gold); but again it might stand for *argentum* (silver), *argentum vivum* (mercury), or much else, including white lead, sulphur, cinnabar, tin, vinegar, saltpeter, and orpiment. See the 67 pp. of entries under "A" in Martinus Rulandus's *Lexicon of Alchemy*, tr. A. E. Waite (1964). No solution appears possible on this basis.

Beginning once again on a totally non-alchemical basis, we might try to find Latin or Greek words beginning with the initials for, let us say, earth, air, and sky, the air being understood to fill the space between the earth and the moon's sphere. The Latin *aer* would do very well for "air" and *stellae* for "stars" (made of fire), but O is an unlikely beginning for any word meaning "earth." In Greek α might stand either for ἀήρ (air) or ἀστέρες (stars), but the omicron and sigma do not appear to fit. Alpha and Omega are ruled out by the physico–chemical context. Hebrew offers another set of possibilities which I am incompetent to explore.

Before abandoning the aphorism, I add general considerations. First, despite the hint in "pyrologians," the aphorism may not relate basically to alchemy. Men skilled in the theory of fire are not restricted to attempts to transmute. Secondly, any set of three letters yields six permutations, not three. Dee does nothing with the orders Ȧ Ṡ Ȯ, Ȯ Ȧ Ṡ, and Ṡ Ȧ Ȯ. For these omissions there must be some reason. Thirdly, it has been seen that "notariacal designation" sometimes has to do with the numerological values of letters. According to Cornelius Agrippa, whom we know Dee to have read with attention, the most probable values are A = 1, O = 50, and S = 90. See *De Occulta Philosophia*, II:xx. The series 1–50–90, 50–90–1, and 90–50–1 is, however, unhelpful.

Finally, what was called the *ars notoria* (not usually *notaria*) was adapted to a variety of concealments, of which I mention two merely to illustrate the range of possibility. In *Ars Notoria: The Notory Art of Salomon* (1657), Robert Turner explains that each of the liberal arts can be acquired by saying a prayer which sometimes is elaborately pious but often consists chiefly of angels' names and nonsense-syllables. In a word, a few minutes of mumbo-jumbo substitute for years of study. Nothing of the sort is likely to be in question here. The *ars notoria* also verged upon the *Steganographia* of Joannes Trithemius (written by 1500; first printed 1606), which encodes messages in prayers and nonsense-syllables, the latter to be addressed, by pretense, to designated angels (whose names were, in fact, keys to the ciphers). It is known that Dee went to pains to obtain a manuscript of the *Steganographia*, but no connection of it with XVIII can be established.

XIX

For the art of graduation, which has to do with mixing substances hot (or cold) and moist (or dry) in varying degrees, see above, pp. 20–21. For example, two measures of something hot in the second degree with one measure of something hot in the fourth degree would give a mixture hot in the two and two-thirds degree.

XX

The elements involve the four qualities (see 1558 reading). Each of the elements has two of the qualities: earth is cold and dry, water cold and moist, air hot and moist, fire hot and dry. Man's "humors" were associated with the elements: phlegm with water, blood with air, and black and red bile with fire. See, again, the title page of Fludd's *De Supernaturali*, etc. (Note XVIII, above). Another figure in the same book (p. 105) correlates fire with the gall bladder, air with the blood of the liver and the veins, water with phlegm and the stomach, and earth with the excrements of the bowels. Dee wants the astrologer to determine what elements each of the "parts, humors, and spirits of the human body" consists of and hence to discover the parts' qualities. For example, because the stomach is associated with water it will be especially affected by the moon, said in CIII to be the chief governess of moist things.

XXI

The implication is that a whole life may be determined, within the limits of free will, by the configuration of the heavens at the time of birth. Such "houses" as those called "Marriage," "Children," and "Death" allowed predictions of future events.

XXII

The primacy given to light again tends to justify belief in astrology, which depends partly upon visible rays streaming from the heavenly bodies.

XXIII

The purpose is both to assert the sensitivity of the whole man to physical influences like those of the heavens and to suggest that music attuned to the harmonies of the spheres is curative for soul and body. The ancient musical modes— Doric, Lydian, etc.—were thought to work almost irresistibly upon the spirit.

XXIIII

The magnet's attractive virtue is perhaps analogous to astrological forces which tend to cause events, its repulsive virtue analogous to astrological forces which tend to prevent or inhibit, and its ability to penetrate bodies with its invisible "rays" analogous to the celestial rays "of more secret influence" which throughout the *Propaedeumata* are distinguished from the sensible rays. The insensible rays are said in XXV to penetrate the earth's thickness. For the magnet's "desire" for a particular orientation to the sky see William Gilbert, *De Magnete* (1600), I: 3, 4.

XXV

The complexity of astrological influences prevented them from being ascribed wholly to visible light; moreover, the traditional philosophy readily imputed importance to invisible entities (e.g., the archetypal ideas of Plato and Aristotle). The attractive and repulsive energy of the magnet, mentioned just above, is a simple example, and its species, too, "penetrated." Penetration was not necessary to assure that the invisible rays operated on every part of the earth's surface; like the visible rays, they might do so as the result of being reflected by the primum mobile (XXVIII).

XXVI

The manifest meaning is that astrological influence is more or less powerful in proportion as the matter exposed to it is more or less receptive, as the design of a seal is more easily impressed upon wax than upon wood or metal. The durability of the impression also varies.

Whether Dee was also thinking of Ficinian talismans (see Note X, above) is uncertain. The mention of *Gamaaeas*, here translated "talismans," suggests that he was: "The object which received the influence and exhibited the sign thereof appears to have been termed Gamaheu, Gamahey, etc. But the name was chiefly given to certain stones on which various and wonderful images and figures of men and animals have been found naturally depicted" (*The Hermetical and Alchemical Writings of Aureolus Philippus Theophrastus Bombast*—i.e., Paracelsus—tr. A. E. Waite, 1894, p. 51n.). If Dee meant the word in its "chief" sense, however, no human operations need be involved.

The problem is best left unresolved until a scholarly consensus is reached about the *Monas Hieroglyphica* (published four years before the second edition of the *Propaedeumata*). An interest in Ficino's talismans certainly continued throughout Dee's lifetime and beyond. See Francis Yates, *passim*, in *Giordano Bruno and the Hermetic Tradition* (1964); also Le Sieur de L'Isle (Charles Sorel), *Des Talismans, ou figures faites sous certaines constellations* (1634).

XXVII

The insensible rays are again meant. The "natural arrangement" or "artful preparation" might refer to the gathering of materials with similar astrological associations—for example, silver, the hyacinth, and the topaz when Jovial influence is sought. Although Dee does not say so, traditionally the materials were to be exposed at an astrologically appropriate time: in the imagined case, on Jupiter's day (Thursday), and in one of the "unequal hours" ruled by Jupiter—the first after sunrise, the eighth, the fifteenth, or the twenty-second. But Dee may be thinking rather of a time when Jupiter is on the meridian, at perigee, etc.

The fact that the "arrangement" or "preparation" is to involve "visible form" as well as "elemental qualities and other properties" again suggests talismans. The incising of an image of Jupiter on a silver plate would suit the conditions; and the influence might then be absorbed by a man who wore the medal. See also Note CX.

XXVIII

"Mirror" is used of the primum mobile—traditionally opaque—to assert that astrological rays from bodies beneath the horizon can be reflected by it. The imperviousness of the primum mobile to astrological rays derived from its solidity and opacity as the boundary of the material universe.

XXIX

The interest shown here in reflected and refracted rays will be picked up in LII, which has to do with the strengthening of light by mirrors. The reflection of the sun's light by the moon is probably also in question; whether by the planets also is uncertain. ("Stars," here and elsewhere, if not modified by "fixed," usually includes the planets, "wandering" being dropped from *stellae errantes*.) See XXXIX, where "the various distances of the stars from the earth" must refer to the planets. The stars were "fixed" in a single sphere.

Whether or not Dee recognized that the planets have little or no intrinsic luminosity is an interesting question. The discovery of the phases of Venus, which pinned the matter down, had to await the invention of the telescope. Cyrano de Bergerac, in his fictive voyage to the sun (c. 1650), noticed that Venus appeared as a crescent and inferred that all the planets depended on the sun for their luminosity as the moon did. See *Voyages to the Moon and the Sun*, tr. Richard Aldington, 1962, p. 191. The comment suggests popular misunderstanding.

XXX

Although *magnitudines* is regularly used in connection with conventional degrees of stellar brightness, Dee is obviously thinking here of bulk.

XXXI

XXXI–XLIV have to do with the intensity of sensible rays in various circumstances. The strength of astrological influence is implied by the phrase "imprint their rays" in XXXVIII. For the geometry relevant to this and the next thirteen aphorisms, see above, pp. 63–66.

XXXII

The rays are, of course, strongest when the celestial body is directly overhead and at perigee. "Star" again means "planet." No single fixed star had significant astrological importance; and even the zodiacal signs were rather conventional "places" than actual constellations.

XXXIII

The cone contains all the rays that converge upon a given external point. The bounding circle of the cone is described by a circle produced by lines from the vertex to all possible tangents with the star.

XXXIIII

Each point of the star within the cone's base sends a ray to the external point. The strength of the rays weakens from the cone's axis toward the perimeter of its base.

XXXV–XXXVI

The aphorisms can be simply illustrated, as follows:

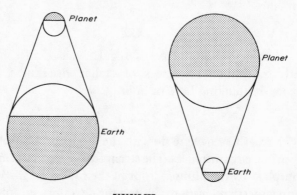

XXXVII

Although the principle that "illumination varies inversely with the square of the distance" had not yet been formulated, Dee of course realized that intensity increases with nearness. That a near planet smaller than the earth

illuminates a smaller part of the earth than a distant one is
shown below.

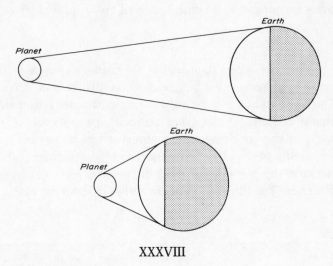

XXXVIII

Again an illustration:

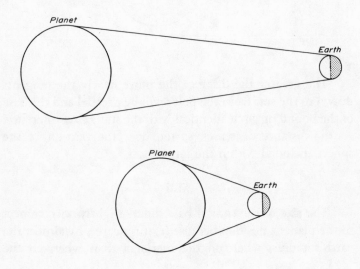

XXXIX

A recapitulation of the preceding four aphorisms. The truncated cones are implicit in the figures above: none of the cones terminates in a point, hence all are truncated.

XL

The base is less than half of the planet's convex surface because tangents must be drawn to the planet from a point. The computation by which the planet's diameter is measured would require (1) the (very difficult) computation of the angle at the cone's apex (the observer's point of view); (2) knowledge of the planet's distance; and (3) adjustment for the fact that the planet's apparent diameter is less than its real diameter. The illustration below shows an extreme case.

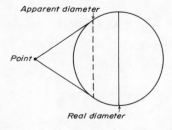

XLI

The greater the distance the more nearly the tangents drawn to the star from the point will be parallel and the base of the bounding circle identical with the star's circumference. As the distance decreases, so too does the portion of the sphere included within the tangents.

XLII

The size of the conical base differs slightly with respect to the planet's position because (1) it is nearer, by almost the earth's radius, when on the meridian than when on the horizon; (2) its orbital distance varies, from a minimum at opposition to a maximum at conjunction (for a superior planet).

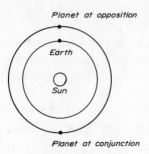

XLIII

All the cones that have been described are "right" cones because they might be produced by the revolution of a right triangle around one of the perpendicular sides. The farther off the radiating planet is, the more nearly all its rays will coincide with the axial ray, which is strongest. Also, the luminous base of a distant planet is greater. "Naturally and simply," however, brilliance increases with nearness. See above, pp. 64–65.

XLIIII

Although all the planets actually have phases, during the sixteenth century only the moon's phases were known. See Note XXIX. A half-moon radiates half as much light as a full moon.

XLV

The true horizon is determined by a plane passing through the earth's center and perpendicular to a diameter dropped from the observer's position. Note the substitution of *mundi* for the *terrae* of 1558: the universe's center is the same as the earth's, hence the universe is geocentric.

The sensible horizon is limited by the observer's line of

sight. It would increase with his elevation above the earth's mean surface. Because of the immense celestial distances, the difference between the two is usually negligible.

As explained above, pp. 73–74, the "Corollary" depends on the fact that three points determine a plane only if they are not collinear. The three points concerned are those of the two points in the sky and the observer's position (or, since the true horizon is in question, the earth's center).

XLVI

This is consequent upon the larger diameter of the star (or planet), thus:

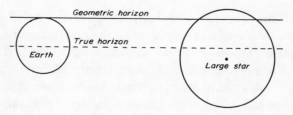

The star in the illustration can be imagined as either rising or setting.

XLVII

See Note XLII: the star is farther when on the horizon than when in the zenith. The sun is nearer to the earth in Capricorn (December) than in Cancer (June) because of the eccentricity of the earth's path about it. Winter is caused not by the sun's greater distance but by the greater obliquity of its rays and the shorter days for people living in the northern hemisphere.

XLVIII

Twilight results from the refraction of the sun's rays by the atmosphere. In the same way, the planets, both superior (farther than the earth from the sun) and inferior (nearer than the earth to the sun), can "communicate the virtue of their accidental light" to us when they are below the line of sight. They emit their species upon the atmosphere, which then, in turn, emits to the observer a species of the species.

XLIX

The reason for praise of God is his use of the air, clouds, water, and mountains to increase the illumination of the earth, to produce visible beauty (as of the rainbow), and perhaps also to strengthen and to complicate astrological influence.

L

Ptolemy—and others after him—thought that gods' names had been given the planets because the planets' "virtues" resembled those popularly ascribed to the gods. Adoption of the belief by Christians permitted them to avoid the imputation of paganism. If Dee meant this, his phraseology was injudicious: "from the imposition of the name of its god." It was a common Christian belief, however, that each planet was ruled by an angel, its "Intelligence." The use of "God" was not restricted even in Christian practice—or in the Bible—to Yaweh or even the Persons of the Trinity. It is unlikely that Dee's educated readers would have been shocked by this aphorism.

LI

This principle was necessary to assure that every birth chart should differ from every other. Antonio Francesco Bonatti, in *Universa astrosophia naturalis* (1687), pp. 15–16, went so far as to claim that not only the birth-moments but also the astrological horizons of twins differed (because of their spatial separation in the womb).

LII

For catoptrics, which had to do with mirrors and reflections, see above, pp. 67–73. Ancient achievements in catoptrics are occasionally alluded to in catalogues of lost skills. In *Epistolae Fratris Rogerii Baconis* (1618—an edition containing notes attributed to Dee), pp. 39-42 are on mirrors and multiplying glasses.

The Monad of the last sentence before the Corollary is reproduced on both title-pages of the *Propaedeumata* and is discussed in a note on them. A printed note in the margin of the 1568 text says, "Ista Insignia fusissime habes explicata, in nostro nuper emisso libello, Cui est Titulus MONAS Hieroglyphica."

LIII

The meaning appears to be that mirrors can be used to focus light (and heat) from the moon during a total eclipse, and the result compared with that obtained from the full moon. The experiment would not have been practicable in Dee's time.

LIIII

The first half of the first sentence means that "Perpendicular rays are stronger than oblique." The first half of the remaining part is explained in Note XLII. The remaining part of the sentence is unclear, but the meaning may be that when the star or planet is perpendicular the rays reflected from the primum mobile coincide approximately with the direct rays.

The remainder of the aphorism is difficult. The eccentricity of the planets' apparent orbits about the earth was established, on the assumption that the planets' speeds were constant, by measurements made from the earth of the differing arcs they described in equal periods of time. From these calculations perigees and apogees could be determined; but I do not understand the phrase about "the angle of incidence."

LV

As when a film is exposed to light for a longer period of time.

LVI

The "powers" are astrological. They are strongest when the planet is at its perigee, on the meridian, and direct (i.e., moving from west to east in the order of the zodiacal signs and thus counterpoising the apparent movement of the heavens).

LVII

This doctrine was necessary to account for the belief that —for instance—planets in the fifth house, "Children," influenced future events.

LVIII

The "equal hours" are the familiar ones, constant in length throughout the year. The unequal hours were computed by divisions into twelve parts each of daylight and darkness. The day-hour and the night-hour are of equal length only at the equinoxes.

LIX

The truth of this observation can be seen by taking extremes: the pole star appears to move very little; points on the celestial equator move very fast.

LX

The apparent speeds of celestial bodies are proportional to the circumferences of the circles they describe parallel to the equator, and the circumferences, in turn, are proportional to the radii or diameters.

LXI

The periods are of several kinds: sidereal (return to the same place relative to the fixed stars); synodic (recurrence of conjunction with the sun), diurnal (return to the meridian), tropical (relative to the equinoctial point), and so on. The return is to a place "very like" the former one because the object or point may have changed in latitude (distance north or south of the ecliptic) by the time it has returned to its original longitude (place among the zodiacal signs).

LXII

Most of the circles will cause no difficulty. The epicycle is a circle whose center is mounted on a rough equivalent of the planetary orbit in such a way that the planet, moving on the epicycle's circumference, appears to an earthly observer to move erratically. Such epicycles were familiar from Ptolemy's time through well beyond Copernicus's. (Properly, the path described by the planet's compound movement is the "orbit," and the circle on whose circumference the epicycle's center moves is the "deferent.") For the technical content of this and the ten following aphorisms, see above, Figure 19 and pp. 80–82.

LXIII

Circles of position "do all cross one another in the North and South Points of the Meridian" (cited in *OED*, art. "position"), like the meridians on the terrestrial globe. Dee suggests that another system of circles can be based on the ecliptic, which cuts the equator at an angle of 23°27′. Still another can be based on an observer's true horizon, which would coincide with one of the others only if his position was at the pole or on the equator.

What is in question is "general ways of drawing horoscopes": specifically, the manner of dividing the zodiac among twelve "houses," upon which horoscopal readings depended heavily. The zodiac was to be cut into twelve segments; but on what circle were the divisions to be based? The choice of the zodiac itself would result in equal segments of 30°. A circle based on the true horizon, or one based on the earth's equator, would result in unequal but differing divisions.

For an account of the several systems of house division, see above, pp. 94–98.

LXIIII

The equatorial period is the time between successive meridian crossings of a fixed star.

LXV

The usual measurement is from noon to noon, that is, between successive meridian crossings of the sun. The length of the natural day varies because of the inclination of the ecliptic and the slight ellipticity of the sun's path. Our day, the civil day, is the average length of the natural day.

LXVI

The tropical year is the time between the sun's return to the same cardinal point, e.g., from the beginning of one spring to the beginning of the next. The actual positions of the constellations vary slightly from year to year because of the precession of the equinoxes. Precession is discussed above, p. 78.

LXVII

The sidereal year—see Note LXI—is slightly longer than the tropical because of the precession.

LXVIII

The return of the moon to the same longitude of the ecliptic would be its sidereal period. To return to the same phase, i.e., to the same position relative to the sun, it must make up the distance the sun has traveled along the zodiac during the moon's sidereal period. The interval between equal phases, the synodic period, is therefore longer than the sidereal.

LXIX

The lunar day is the time between the moon's successive crossings of the meridian. Because of the moon's irregularities —caused by the gravitational pull on it of both earth and

sun—the lunar day is not constant. The slowness of apparent
movement in the fixed stars (the precession) is such that a
complete revolution of the stars around the earth requires
about 26,000 years.

LXX

The periods are the equatorial (LXIIII), diurnal (LXV),
tropical (LXVI), and sidereal (LXVII). Despite the compli-
cations caused by the epicycles, the astronomer "should
know" (LXVIII) the periods not only because accurate knowledge is valuable for its own sake but also because the planetary positions are important in astrology.

LXXI

The passing of the moon across (or beneath) the sun's
face in conjunction takes some minutes, so that the time of
exact conjunction requires computation. The sun moves
more slowly (against the stars) than the moon, as—for instance
—Mars moves more slowly than Venus. In consequence, the
next conjunction occurs before the slower body has completed
a full revolution.

LXXII

The diurnal period of a planet (its return to the meridian)
is much less than its synodic or sidereal period.

LXXIII

The gist is that careful observation will reveal connections
between earthly objects or happenings and celestial phenomena. Thus the dominance of Capricorn might quickly be
associated with cold winds, and that of Leo with hot weather.
Smaller-scale happenings would be harder to relate to cosmic
events: for example, the dismal failure of a political address
to retrograde movement of Mercury, the Lord of Language,
in the third house, which although called *Fratres* related also
to speech. Nevertheless the claim was regularly made that
astrological predictions had been made possible by several

thousand years of "experience," or empirical investigation, begun by the Chaldaeans (the Assyro-Babylonians) and continued ever since. Dee would almost certainly have claimed that observation had proved the association of the planets with specific metals. How far the system could be pushed is evident in Ficino's association with planets of plants, stones, animals, odors, clothing, and even dances and songs. Cf. Shumaker, *Occult Sciences*, pp. 120–133.

In Inference 1 the resemblance of sublunary objects and events is said to produce "harmony." The whole concept has been described by E. M. W. Tillyard in *The Elizabethan World Picture* (1950), Chap. 6, in terms of "corresponding planes," and by Heninger, *Touches of Sweet Harmony* (1974), in Pythagorean terms. The three levels of Inference 1 are explained in Inference 2 as the celestial, the terrestrial, and the microcosmic or human and are illustrated by the sun, gold, and man's heart, each a "primate" in its own frame of reference.

LXXIIII

"Some place in the sky" would include such geometrical points as the Lot of Fortune and the moon's nodes. The significators are, primarily, bodies of particular importance for marriage, longevity, etc. The general meaning is that the influence of a luminary is affected by its position in space and its relations to signs and other planets. In normal astrology, the angular relations of planets with one another ("aspects") were especially important. A trine relationship (roughly 120°) was "good," a square (90°) "bad," etc.

LXXV

The apparent persistence of the fixed stars in their relationships to one another implies superiority. Against this background the planets, although because of the earth's turning they rise in the east and set in the west, move eastward. The precession, however, which slowly alters the position of the stars relative to the equinoctial points, suggests that even our most important earthly affairs—those which resemble the stars in having special eminence—suffer vicissitudes. The reasoning is by analogy. The harmony that binds the fixed stars in their relative positions also binds them to the elemental kingdom, which stretches from the earth to the moon. Finally, radiation is the instrument of all this binding.

LXXVI

The slowness of the precession implies a slow development of stellar influence. Interpretation of the influence is complicated by the fact that rays emanate simultaneously from all the stars, so that the effect of a single one, or even of a whole constellation, is hard to isolate.

LXXVII

The gist is that an apparently feeble planet or constellation can sometimes exert more influence than an apparently stronger one either because the subject is naturally related to the luminary—as silver to the moon—or because artful preparation has made it especially susceptible to influence.

The remainder of the aphorism is rather associative than logical. Dee gives an alchemical parallel (a gentle agent may be more useful than a strong one) and then adds a remark that is relevant to the alchemical context but not to the astrological one. The Holy Art (alchemy) requires repeated distillations or purifications followed by repeated operations to reunite ("marry") what has been separated, as is suggested by a Scriptural text. The key to "David's Seven Times" is given by the Hebrew characters for "12" in the margin. *Psalm* 12:6 reads, "The words of the Lord are pure words: as silver tried in a furnace of earth, purified seven times." *Dualiter* refers to the dual form of "Seven Times." Dee interprets the text to mean that the seven purgative operations are to be followed by seven compoundings. (I owe this reading of the Hebrew to my colleague John Coolidge.) It would be interesting to know whether Dee, who believed strongly in numerology, noticed that the number of this aphorism is 77, or perhaps even placed the aphorism in this position deliberately. One guess might be that observation of the number reminded him of *Psalm* 12:6 and led him to the alchemical digression.

LXXVIII

In fact the stars are enormously larger than was thought in Dee's time—or in Ptolemy's. The nearest, our sun, has a diameter 108 times the earth's and a volume 108^3, or 1,159,712, times greater. Yet the sun is a dwarf star. For the ancient technique of estimating star sizes, see above, pp. 87–88.

LXXIX

For a natural day, see LXV. It is four minutes longer on the average than the equatorial period. The difference will define the daily western movement against the sun of a point on the equator ("diurnal direction").

LXXX

We are now asked to find the diurnal directions of other circles.

LXXXI

Diurnal direction can be measured in the same way for a lunar day as for a solar one. The two results would differ because the "days" are of unequal lengths.

LXXXII

A planet's or fixed star's "day" is a "diurnal horizontal period": the time between successive risings.

LXXXIII

The purpose is again that of determining differences in "periods."

LXXXIIII

Why the powers of the sun and moon should be made "very clear" by the calculations just recommended is obscure. Possibly the idea is that the periods determine the *mora*, or "delay," of the body between the horizons, hence the length of time its rays are direct. The last two sentences appear to mean that observational data are more reliable than conventional approximations.

LXXXV

A planet's movement is retrograde when it appears to move from east to west in its private motion. See above, pp. 80–82. The result is a "pushing back" of celestial points, or "reversed direction," since the planet, by moving in the same direction as the equator, has a diurnal period less than its equatorial period.

LXXXVI

The sidereal period of Jupiter is 11.86 years; consequently in one year its position relative to the fixed stars will have altered by about 30°. *Dodecatemorion* comes from Greek roots meaning "twelfth piece"; here it means one zodiacal sign (30°).

The degree of displacement differs from year to year because Jupiter's orbit is based not upon the earth but upon the sun. The last sentence appears to mean, "Take a lot of such measurements, not just a few." Other planets' movements are to be measured in the same way.

LXXXVII

The equator's harmonic period is that in which a point on it makes a single complete revolution. Planets moving in a direct course pause longest between the horizons, hence irradiate exposed objects more effectively. Their "benevolence," however, is the contrary of the recalcitrance implied by retrogression.

LXXXVIII

Further exploration of problems implicit in retrograde motion, which because the planet is running counter to its dominant course reminds Dee of an uncontrolled horse, a subordinate part of nature rebelling against a superior part, a discordant note, a violent agent that causes damage, and a force that willfully negates itself. The "most excellent" of the planets, the sun and moon, are never guilty of these misdemeanors, and the others commit them only briefly. The anomalies are justified on the ground that their result is a more pleasing ornamentation of nature. Cf. XII. Dee infers that during retrogression a planet's influence is weakened (partly because of its shorter *mora*) or reversed.

It will be observed that "fourthly" is omitted, apparently by inadvertence.

LXXXIX

So far as "proper signification" is concerned (see LXXIIII), planetary powers are greatest when the planet is furthest distant. The base of its radiant cone is then largest, and all its rays are angularly nearer the axial ray, which is strongest. See XXXIIII. In matters other than proper signification, strength increases with nearness because of the increase of intensity.

The passage about making a planet stand off very far from the earth may or may not refer to visual appearance, as when an object is looked at through the wrong end of a telescope. For a detailed discussion see above, pp. 68–70. Repulsion may require "a few days" for the preparation and arrangement of the mirrors, whereas attraction can be produced by changed adjustments in mirrors already at hand.

The quoted phrase at the end is from *Acts* 4:25, which itself adapts the Septuagint text of *Psalm* 2:1. The meaning is that the operation might cause popular resentment by awakening fear.

XC

The usual astrological view was that a planet lost its powers when it was combust (so near the sun as to be "burned"). The allusion to the rules of graduation suggests that the effect of combustion on a given planet can be computed by the process commented on in Note XIX. Nearness to the sun may increase the heat of a "hot" planet and thus strengthen its influence.

XCI

The earth's central position (geocentrism) in the cosmos is privileged by Divine ordinance: another indication of Dee's rejection, at least as late as 1568, of the Copernican hypothesis. That he knew Copernicus's *De Revolutionibus* is clear from LXVII. Cf. above, pp. 56–57.

XCII

Any two stars or planets exactly opposite each other have equal opportunities to exert influence because, as the heavens turn, they will gradually exchange places along the same path. The "working together" of the stars might be either complementary or antagonistic. "Declination" is the angular distance above (or below) the celestial equator. The "schematic or aschematic intervals" may be those of exact or approximate opposition, the declinations and distances from the meridian being identical or only roughly equivalent.

XCIII

At all places except the poles and the equator the horizon cuts the sphere of the stars "obliquely": that is, the line drawn to the observer's zenith makes an angle other than 0°

or 90° with the world's axis. In such places the segments of great circles cut off between the horizon and the meridian are usually not arcs of 90° (quarter-circles). The qualification is for great circles, like the celestial equator, that pass through the east and west points of the horizon. Hence the ninetieth degree of the ecliptic west of the ascendant falls on the meridian only when an equinox is rising. Consequently, the intersection of the ecliptic and the meridian, which marks the cusp of the tenth house, coincides only rarely with the ninetieth degree or the highest point of the ecliptic above the horizon, the so-called Heart of the Sky. Cf. above, pp. 95–96.

XCIIII

This theoretical—though useless—truth must have been inferred from the obvious heat of the sun's rays, light from a candle, etc. Even the heat in the moon's light can hardly have been measurable in Dee's time.

XCV

For the sun's size, see Note LXXVIII. But of course the sun is by no means the largest of the stars.

XCVI

The scale Dee is using evidently is arbitrary: 100° or 60° = "maximal." At perigee, the sun's rays fall obliquely upon the northern hemisphere. The situation he describes could exist only on the Tropic of Capricorn.

XCVII

The points over which the sun can stand perpendicularly are between the tropics.

XCVIII

Evidently the temperature of a substance exposed to the sun is to be measured; but how? Thermometers were a development of the seventeenth and eighteenth centuries.

"In actual fact" (literally, "according to the truth of the matter"), opposed as the phrase is to mere proportionality, suggests rather experimental knowledge than computational. Nevertheless Dee must have been thinking here, as elsewhere, of calculations.

XCIX

The "different places" of the sun may differ latitudinally —but only if "perpendicular" means "on the meridian" and not "in the zenith." If the latter, varying heat would be produced, in theory, only by differences in elevation, as between a mountain top and a valley, or by differences in the sun's distance, as at perigee or apogee. The "artifice" is presumably catoptrical: by a proper placing of mirrors or shields the heat could be increased or reduced in any desired proportion.

C

Taken literally, the first sentence would again imply that Dee did not know that the planets have no, or very little, luminosity. See Note XXIX. Here the difference between the sun's heat and a planet's is said to derive merely from "bases" and "distances." The third sentence acknowledges the admixture of some special "sensible quality"—probably moisture, dryness, etc.—with heat in each of the planets and recommends that this be determined by contrasts between the planet's effects on some exposed body during periods of visibility and invisibility.

CI

The process just described for determining the sun's heat is now to be applied to the moon. We are warned, however, that account must be taken of the moon's phases. When the moon is horned the "vertical axis" described in XXXIIII as strongest is not operative.

CII

The astrological influence of the sun benefits especially from the quantity of its rays, that of the moon from the greater swiftness of its motion, believed also to contribute to astrological influence. Relatively to the stars, the sun moves eastward about 1° a day and the moon about 13°.

CIII

Historically, the moon has always been considered "moist." For example, the sap in plants was thought to increase with the degree of the moon's fullness.

CIIII

Rapid motion implies energy, vigor. The sun's dominion over "vital heat" is here imputed to its light, the moon's control over moisture to its movement: which would also, however, determine its *mora*. See the next note.

CV

The nearness of the moon is considered here in relation not to the intensity of its light but to the speed of its motion. A body moving in an eccentric orbit appears to move fastest at perigee, slowest at apogee.

CVI

Roughly, the sun's rays are the source of heat, the moon's movement the source of moisture. XCV–XCVIII have been about the sun's heat, CIII–CV about the moon's movement.

CVII

The comparison of the seasons to a day (and of both to a lifetime) was conventional. The sun's heat could "produce" two of the primary qualities, cold and moisture, only by privation, i.e., by not working. The "necessary order" might be some such schema as the following: morning, hot and moist; afternoon, hot and dry; evening, cold and dry; night, cold and moist. The "necessary order" requires that only one quality at a time be replaced by its opposite.

The Duodenary or twelve apparently results from the combinations of beginnings, middles, and ends with either the seasons or the four parts of the natural day. At the terrestrial pole the seasons and the parts of the day would coincide. How the Trinity is involved is obscure; but one occasionally reads, in the period, that the "beginning" (from the creation to the birth of Christ) was the period of the Father, the "middle" (to Christ's Second Coming) is the Age of the Son, and the "end" (from the Second Coming onward) will be the Age of the Holy Spirit. Those who pant after concealment would be occult philosophers who veiled their writings to hide mysteries from the unfit.

Twelve, together with its factors three and four, had especially rich numerological significance. Of the many expositions of this tiresome subject one of the more detailed is Peter Bongus's *Numerorum Mysteria, Ex abditis plurimarum disciplinarum fontibus hausta* (1618).

CVIII

The twenty-six positions of the sun with relation to the constellations appear in the *Almagest*, VIII:iv. Ptolemy divides these into nine general classes subdivided into parts which total twenty-six.

"Early morning east wind" has to do with the nearly simultaneous co-rising of the sun and a given star in the constellation. The star may rise a little later ("invisible after-rising"), at exactly the same time ("true morning co-rising"), or slightly earlier ("visible morning co-rising"). Note that the general name, "Early morning east wind," does not count toward the total of twenty-six. And so on.

Dee urges that a similar set of twenty-six positions should be distinguished for each of the seven planets (including the sun). The 182 different reckonings would result from the multiplication of 26 by 7. The result would be nothing more than the enrichment of the technical vocabulary.

CIX

The point is that because the death of the body results from physical causes the astrologer can predict a natural limit for a specific human life. If death occurs before the natural term, the cause is human "negligence." The basic doctrine is partly contradicted by the close association between soul and body asserted in XXIII; but the discrepancy could be removed by a simple qualification.

CX

In scholastic philosophy, "form" was "The essential determinant of a thing; that which makes anything (*matter*) a determinate species or kind of being ... " (*OED*). Since immateriality is of a higher order than matter, its powers were superior. The "visible form" of talismans discussed in Note XXVII may relate obscurely to the prestige of "form."

CXI

The insensible or invisible rays are intelligible because their existence has been established by mental operations. Astrological knowledge was supposed to be based on several millenia of observations and inferences, as remarked in Note LXXIII. The analogy of the insensible rays with the invisible soul imputes dignity to them.

CXII

Cf. the assertion by Mantice, in Pontus de Tyard's *Mantice, ou Discours de la verité de divination par astrologie* (2d ed., 1573), pp. 91–92, that the majesty of Saturn, the benignity of Venus, and the warmth of Mars are not naturally maleficent but are "worsened" in some human beings.

CXIII

Different parcels of matter have differing potentialities for reaction to planetary and stellar radiation. Presumably further diversity is produced by differing astrological horizons. No two spots on the earth's surface are bathed in exactly the same strength by visible and invisible rays.

CXIIII

The elemental part of the cosmos lies within the moon's sphere. Above that is the empyrean, the realm of pure fire. Every existent object apparently is either produced by the celestial harmonies or itself possesses them (or both). Sometimes the relationship is obvious, sometimes obscure.

CXV

The relationships mentioned in the foregoing aphorism should be compared and analyzed for the improvement of astrological science.

CXVI

This and the following aphorism are analyzed in detail above, pp. 90–91. Dee was fond of playing with permutations and, within limits, skilled in doing so. Since the planets' orbits do not lie in exactly the same plane, many of the conjunctions would not be linear. The utility of the computations for practical astrology—for which, by reason of the emphasis on comparative "strengths," they seem to have been intended—is doubtful, or worse. The strengths taken into consideration are merely "greater," "equal," and "less."

CXVII

The "conjunctions" now become merely those of converging rays. The fact that intrinsically different strengths can become identical depends on angles of incidence and positions in houses as well as upon orbital variations: in a favorable sign a planet is "exalted," in an unfavorable one "dejected." The calculations, we are told, can be extended almost to infinity (by sophisticating the degrees of strength). Although Dee is showing off here, as often, he is saying implicitly that popular astrology is culpably inexact.

CXVIII

The gist is an affirmation that efforts should be made both to localize the effects of understood astronomical events and, contrariwise, by examining remarkable earthly happenings to determine the significance of hitherto uninterpreted celestial configurations. In this way astrological knowledge can be increased.

Matthew 2:2, quoted in the "Comment," was often used to legitimize astrology.

CXIX

Hermes Trismegistus, "thrice-great" as greatest philosopher, king, and priest, was regarded as the founder of both astrology and alchemy. He was supposed to have been roughly contemporary with Moses. His vogue in the Renaissance began with Ficino's translation of the *Pimander* (now known as the *Corpus Hermeticum*) in 1461. For a summary of Hermes's doctrines and influence see Shumaker, *The Occult Sciences*, Chap. V.

CXX

A Greek authority (Ptolemy? I have not identified the sentence) is adduced to corroborate the Egyptian Hermes and to suggest that Dee has the support of prestigious ancients generally. The emphasis here is on a divinely instituted "revolution," like that of the heavens, which preserves the continuity of created nature. This introduces a final pious exclamation.

⅏ Bibliography of Works Cited

Only works cited in short form in the notes are listed; the bibliography omits a few titles merely mentioned, and cited in full, in the text.

The bibliography is arranged by author, who might be either a person or an institution. In all cases the author as given in the notes should lead directly to the appropriate item in the bibliography. When the item is one of a collection of essays, it will be entered under its author as usual, but publishing details might come under the name of its editor. For example, the notes cite Vogl in *Bacon Essays*, ed. Little; the title and pagination of Vogl's essay appear in the bibliography under "Vogl," the place and date of publication, and the full title of the *Essays*, under "Little."

Abat, Bonaventure. *Amusemens philosophiques sur diverses parties des sciences, et principalement de la physique et des mathematiques.* Amsterdam, 1763.

Abulfeda. *La Géographie d'Abulféda.* 2 vols. J. T. Reinaud and Stanislas Guyard, ed. and tr. Paris, 1848–1883.

Agrippa, Henricus Cornelius, of Nettesheim. *De occulta philosophia libri tres.* Karl Anton Novotny, ed. Graz, Austria, 1967.

———. *The Vanity of Arts and Sciences.* London, 1676. (First edition, Antwerp, 1530.)

Alfraganus. *Il "Libro dell'aggregazione delle stelle."* Romeo Campani, ed. Florence, 1910. (Collezione di Opuscoli Danteschi Inediti o Rari, vols. 87–90.)

Alessio, Franco. *Mito e scienza in Ruggero Bacone.* Milan, 1957.

Allen, Phyllis. "Scientific Studies in the English Universities of the Seventeenth Century." *Journal of the History of Ideas* 10 (1949): 219–253.

d'Alvernay, Marie-Thérèse, and Françoise Hudry. "Al Kindi, *De radiis*." *Archives d'histoire doctrinale et littéraire du moyen age* 41 (1974): 139–267.

Anonymous. "On Catoptrical and Dioptrical Instruments of the Antients." *Philosophical Magazine* 19 (1804): 176–190.

Archibald, R. C. "The First Translation of Euclid's Elements into English and its Source." *American Mathematical Monthly* 57 (1950): 443–450.

Averdunk, Heinrich, and J. Müller-Reinhard. *Gerhard Mercator und die Geographen unter seinen Nachkommen.* Gotha, 1914. (*Petermanns Mitteilungen* 39.)

Bacon, Roger. *Epistolae Fratris Rogerii Baconis.* Hamburg, 1618.

———. *Frier Bacon his Discovery of the Miracles of Art, Nature, and Magick.* T. M., tr. London, 1659.

———. *Opera hactenus inedita.* R. Steele et al., eds. Oxford, 1909–1940.

———. *Opus maius.* J. H. Bridges, ed. 2 vols. London, 1900.

———. *The Opus Maius of Roger Bacon.* Robert Belle Burke, tr. 2 vols. Philadelphia, 1928.

———. *Roger Bacon's Letter Concerning the Marvelous Power of Art and of Nature and Concerning the Nullity of Magic* [*De secretis operibus artis et naturae, et de nullitate magiae*]. Tenney L. Davis, tr. and ed. Easton, Pa., 1923.

Baxandall, David. "Early Telescopes in the Science Museum, from an Historical Standpoint." London, Optical Society. *Transactions* 24 (1923): 304–320.

Benjamin, Francis S., Jr., and G. J. Toomer. *Campanus of Novara and Medieval Planetary Theory.* Madison, 1971.

Bergerac, Cyrano de. *Voyages to the Moon and the Sun.* Richard Aldington, tr. New York, 1962.

Billingsley, H. *See* Euclid. *The Elements of Geometrie.*

Birkenmajer, A. "Le Commentaire inédit d'Erasme Reinhold sur la *De Revolutionibus* de Nicolas Copernic." (Colloque de Royaumont, 1957.) *La Science au Seizième Siècle.* Paris, 1960. Pp. 169–180. (*Histoire de la Pensée* 2.)

Björnbo, Axel A., and Sebastian Vogl. "Alkindi, Tideus und Pseudo-Euklid. Drei optische Werke." *Abhandlungen zur Geschichte der mathematischen Naturwissenschaften* 26:3 (1912): 1–176.

Bold, Franz, Carl Bezold, and Wilhelm Gundel. *Sternglaube und Sterndeutung.* 5th ed., Darmstadt, 1966.

Bongus, Petrus. *Numerorum mysteria, Ex abditis plurimarum discipli-narum fontibus hausta.* Paris, 1618.

Bonatti, Antonio Francesco. *Universa astrosophia naturalis.* Padua, 1687.

Book of Formation, The. See Joseph, Rabbi Akiba ben.

Borde, Andrew. *The Pryncyples of Astronomy.* London, c. 1542. (Re-printed, Amsterdam, 1973.)

Bosmans, H. ["Notes que Gemma Frisius a écrites sur les marges de son exemplaire de l'*Arithmetica integra* de Stifel."] Brussels. Société Scientifique. *Annales* 30:1 (1906): 165–168.

———. "Sur le *Libro de Algebra* de Pedro Nuñez." *Bibliotheca Mathematica* 3 (1908): 154–169.

Bouché-Leclerc, A. *L'Astrologie grecque.* Paris, 1899.

Bourne, William. *A Regiment for the Sea and Other Writings on Naviga-tion.* E. G. R. Taylor, ed. Cambridge, 1963. (Hakluyt Society, Second Series, no. 121.)

———. "A Treatise on the Properties and Qualities of Glasses for Optical Purposes, According to the Making Polishing and Grinding of them." *Rara mathematica.* J. O. Halliwell-Phillipps, ed. 2nd ed., pp. 32–47. London, 1841.

Bowden, Mary Ellen. *The Scientific Revolution in Astrology: The English Reformers, 1558–1686.* Unpublished Ph.D. thesis. Yale, 1974.

Brahe, Tycho. *Astronomiae instauratae progymnasmata.* Uraniburg, 1602. (Reprinted in Brahe. *Opera omnia,* J. L. E. Dreyer, ed. Vols. 2–3. Copenhagen, 1915–1916.)

Bridges, John Henry. *The Life and Work of Roger Bacon. An Introduction to the Opus Maius.* London, 1914.

Brussels. Université libre. *L'Univers à la renaissance: Microcosme et macrocosme.* Brussels and Paris, 1970.

Burckhardt, Jacob. *The Civilization of the Renaissance in Italy.* 2 vols. New York, 1958.

Burgon, John William. *The Life and Times of Sir Thomas Gresham.* London, 1839.

Burton, Robert. *The Anatomy of Melancholy.* F. Dell and P. Jordan-Smith, eds. New York, 1927.

Burtt, E. A. *The Metaphysical Foundations of Modern Physical Science.* 2nd ed., 1932. (Reprinted, New York, 1955).

Cajori, Florian. *A History of Mathematical Notations.* 2 vols. La Salle, Ill., 1928–1929.

Calder, I. R. F. *John Dee Studied as an English Neo-Platonist.* 2 vols. Un-published Ph.D. thesis. University of London, 1952.

Cardano, Girolamo. *Aphorismorum astronomicorum segmenta septem.* In Cardano. *Libelli quinque.* Nürnberg, 1547. Ff. 207^r–309^v (*Opera* 5: 29–92.)

———. *The Book of My Life (De vita propria liber)*. Jean Stoner, tr. New York, 1962.

———. *Opera omnia*. Carolus Sponius, ed. 10 vols. Lyon, 1663.

Caspar, Max. *Kepler*. C. Doris Hellman, tr. London and New York, 1959.

Cassirer, Ernst. "Mathematical Mysticism and Mathematical Science." In *Galileo, Man of Science*. Ernan McMullin, ed., pp. 338–351. New York and London, 1967.

Cassirer, Ernst *et al.*, eds. *The Renaissance Philosophy of Man*. Chicago and London, 1948.

Cavellat, Gulielmus. *Annuli astronomici . . . usus, ex variis authoribus*. Paris, 1558.

Clarke, Frances Marguerite. "New Light on Robert Recorde." *Isis* 8 (1926): 50–70.

Clulee, N. H. "John Dee's Mathematics and the Grading of Compound Qualities." *Ambix* 18 (1971): 178–211.

———. *The Glas of Creation. Renaissance Mathematicism and Natural Philosophy in the Work of John Dee*. Unpublished Ph.D. thesis. University of Chicago, 1973.

Copernicus, Nicholas. *De revolutionibus orbium coelestium*. Nürnberg, 1543. (English translation by Charles Glenn Wallis in *Great Books of the Western World*, vol. 16., pp. 501–838. Chicago, 1952.)

Cooper, Charles Henry, and Thomas Cooper. *Athenae Cantabrigienses*. Vol. 2. Cambridge, 1861.

Cortesão, A. *Cartografia e cartografos portugueses dos secolos XV e XVI*. 2 vols. Lisbon, 1935.

Crombie, Alistair C. *Robert Grosseteste and the Origins of Experimental Science, 1100–1700*. Oxford, 1953. (Reprinted, with additions, 1962.)

Daujat, Jean. *Origines et formation de la théorie des phénomènes électriques et magnétiques*. 3 vols. Paris, 1945.

Debus, Allen G. *The English Paracelsians*. London, 1965.

———. "Introduction." John Dee. *Mathematicall Preface* (1970). Pp. 1–33.

Dee, John. *Autobiographical Tracts of Dr. John Dee, Warden of the College of Manchester*. James Crossley, ed. Manchester, 1851. (Chetham Society Publications 24.)

———. "Epistola." *See* Feild, John. *Ephemeris*.

———. *The Mathematicall Preface to the Elements of Geometrie of Euclid of Megara* (1570). New York, 1975.

———. "Mathematicall Preface." *See* Euclid. *The Elements of Geometrie*. ed. Billingsley.

———. *Monas hieroglyphica.* Antwerp, 1564. (Translated into English by C. H. Josten, *Ambix* 12 (1964): 84–221, and by J. W. Hamilton-Jones, London, 1947; into French by Grillot de Givry, Paris, 1925.)

———. *Parallaticae commentationis praxeosque nucleus quidam.* London, 1573.

———. *The Private Diary of Dr. John Dee and the Catalogue of his Library of Manuscripts.* James Orchard Halliwell, ed. London, 1842. (Camden Society Series, no. 19.)

———. *A True and Faithful Relation of what Passed for Many Yeers between Dr. John Dee . . . and Some Spirits.* Meric Casaubon, ed. London, 1659.

De L'Isle, Le Sieur (Charles Sorel). *Des Talismans, ou figures faites sous certaines constellations.* Paris, 1634.

della Porta, Giambattista. *Magiae naturalis libri XX.* Naples, 1589. (Translated and reprinted as *Natural Magick.* London, 1658.)

Dieterich, Albrecht. *Eine Mithrasliturgie erläutert.* Leipzig, 1903.

Digges, Leonard. *A Prognosticon of Right Good Effect.* London, 1555. (Old Ashmolean Reprints, 3.)

Drake, Stillman. *Discoveries and Opinions of Galileo.* New York, 1957.

———. "Galileo's Discovery of the Laws of Free Fall." *Scientific American* 228:5 (1973): 84–92.

———. "Galileo Gleanings, XXII: Galileo's Experimental Confirmation of Horizontal Inertia: Unpublished Manuscripts." *Isis* 64 (1973): 291–305.

Dreyer, J. L. E. *A History of Astronomy from Thales to Kepler.* 2nd ed., New York, 1953.

Duhem, Pierre. *Le Système du monde.* Vol. II. 2nd ed., Paris, 1965.

Durme, M. van. *Correspondance mercatorienne.* Antwerp, 1959.

Easton, Joy B. "The Early Editions of Robert Recorde's *Ground of Artes.*" *Isis* 58 (1967): 515–532.

———. "A Tudor Euclid." *Scripta Mathematica* 27 (1966): 339–355.

Elton, G. R. *England under the Tudors.* 3rd ed., London, 1969.

Essen, L. van der. "Le Rôle de l'Université de Louvain au XVI siècle." *Revue générale belge* 1:43 (1949): 38–64.

Euclid. *The Elements of Geometrie of the Most Auncient Philosopher Euclid of Megara . . . With a very fruitful preface made by M. I. Dee.* H. Billingsley, ed. and tr. London, 1570.

———. *Elements of Geometry.* John Leeke and George Serle, eds. London, 1661.

———. *Euclidis elementorum liber decimus.* Petrus Montaureus, tr. and ed. Paris, 1551.

———. *See* Thomas-Stanford, C. *Early editions.*

Feild, John. *Ephemeris anni 1557 ... Adiecta est etiam brevis quaedam epistola Joannis Dee, qua vulgares istos ephemeridum fictores merito reprehendit.* London, 1556.

Ficino, Marsilio. *De vita triplici.* In *Opera.* Basel, 1576.

Finé, Oronce. *In sex priores libros geometricorum elementorum ... demonstrationes.* Paris, 1536.

Firpo, Luigi. "John Dee, scienziato, negromante e avventuriero." *Rinascimento* 3 (1952): 25-84.

Fludd, Robert. *De supernaturali, naturali, praeternaturali et contranaturali microcosmi historia.* Oppenheim, 1619.

Fracastoro, Girolamo. *De sympathia & antipathia rerum.* Lyons, 1550.

French, Peter J. *John Dee. The World of an Elizabethan Magus.* London, 1972.

Galilei, Galileo. *Dialogue Concerning the Two Chief World Systems.* Stillman Drake, tr. Berkeley and Los Angeles, 1953.

———. *On Motion and Mechanics.* I. E. Drabkin and S. Drake, eds. Madison, 1960.

Geymonat, Ludovico. *Galileo Galilei.* Stillman Drake, tr. New York, 1965.

Gilbert, Ph. "Les sciences exactes dans l'ancienne Université de Louvain." *Revue des questions scientifiques* 16 (1884): 438-453. Reprinted, with notes by B. Lefebvre, *ibid.* 12 (1927): 17-47.

Gilbert, William. *De magnete.* London, 1600.

Giles of Rome. *Errores philosophorum.* Josef Koch, ed., and John O. Riedl, tr. Milwaukee, 1944.

Ginsburg, Jekuthiel. "Rabbi ben Ezra on Permutations and Combinations." *Mathematics Teacher* 15 (1922): 347-356.

Gogava, Antonio. *See* Ptolemy. *Opus quadripartitum.*

Goldstein, B. R. "The Arabic Version of Ptolemy's *Planetary Hypotheses.*" Philadelphia. American Philosophical Society. *Transactions* 57 (1967): 3-55.

Gomes Teixeira, Francisco. *História das matemáticas em Portugal.* Lisbon, 1934. (Academia da Ciências de Lisboa. Biblioteca de Altos Estudos.)

Gunther, R. T. *Early Science in Oxford.* Vol. 2: *Astronomy.* Oxford, 1923. (Reprinted, London, 1967.)

Haardt, Robert. "The Globe of Gemma Frisius." *Imago Mundi* 9 (1952): 109-110.

Halliwell-Phillipps, James Orchard. *Rara Mathematica.* 2nd ed., London, 1841.

Harrington, James. *Oceana.* S. B. Liljegren. Lund, 1924.

Hartner, Willy. "Medieval Views on Cosmic Dimensions and

Ptolemy's *Kitāb al-Manshūrāt.*" *Mélanges Alexandre Koyré.* Vol. 2, pp. 254–282. Paris, 1964.

Heath, Thomas Little. *The Thirteen Books of Euclid's Elements.* 3 vols. 2nd ed., Cambridge, 1925.

Heilbron, J. L. *A History of Electricity.* Berkeley, forthcoming.

Hellyer, Brian, and Heather Hellyer. "King Edward VI's Defence of Astronomy." *British Astronomical Association. Journal* 82 (1972): 362–366.

Henderson, Janice. "Erasmus Reinhold's Determination of the Distance of the Sun from the Earth." In *The Copernican Achievement.* R. S. Westman, ed. Pp. 108–129.

Heninger, S. K., Jr. *Touches of Sweet Harmony: Pythagorean Cosmology and Renaissance Poetics.* San Marino, California, 1974.

Hofmann, Joseph E. *Michael Stifel, 1487?–1567. Leben, Wirken und Bedeutung für die Mathematik seiner Zeit.* Wiesbaden, 1968. (Sudhoffs Archiv. *Beihefte* 9.)

James, M. R. *List of Manuscripts Formerly Owned by Dr. John Dee.* London, 1921. (London. Bibliographical Society. Supplement to the *Transactions* 1.)

Johnson, F. R. *Astronomical Thought in Renaissance England.* Baltimore, 1937.

Johnson, F. R., and Sanford V. Larkey. "Thomas Digges, the Copernican System, and the Idea of the Infinity of the Universe in 1576." *Huntington Library Bulletin* no. 5 (1934): 69–117.

————. "Robert Recorde's Mathematical Teaching and the Anti-Aristotelian Movement." *Huntington Library Bulletin* no. 7 (1935): 59–87.

Joseph, Rabbi Akiba ben. *The Book of Formation.* Knut Stenring, tr. New York, 1970.

Kaplan, Edward. *Robert Recorde (c. 1510–1558): Studies in the Life and Work of a Tudor Scientist.* Unpublished Ph.D. thesis. New York University, 1960.

Keuning, Johannes. "The History of Geographical Map Projections until 1600." *Imago Mundi* 12 (1955): 1–25.

Koch, Walter A., and F. Zanziger. *Regiomontanus und das Häusersystem des Geburtsortes.* Göppingen, 1960.

Kocher, Paul H. *Science and Religion in Elizabethan England.* San Marino, California, 1953.

Koyré, Alexandre. *Études galiléennes.* 3 vols. Paris, 1939.

————. "Galileo and Plato." *Journal of the History of Ideas* 4 (1943) 400–428.

Kristeller, P. O. *The Philosophy of Marsilio Ficino.* V. Conant, tr. New York, 1943.

Kyewski, Bruno. "Über die Mercatorprojektion." *Duisburger Forschungen* 6 (1962): 115–130.

Lange, Gerson. *Sefer Maassei choscheb. Die Praxis des Rechners. Ein hebräisch-arithmetisches Werk des Levi ben Gerschom aus dem Jahre 1321*. Frankfurt-am-Main, 1909.

Langlois, Charles Victoire. "Études sur deux cartes d'Oronce Finé de 1531 et 1536." Paris. Société des Américanistes de Paris. *Journal* 15 (1923): 83–97.

Libavius, Andreas. *Alchemia . . . In integrum corpus redacta*. Frankfurt, 1597.

Lindberg, David C. "Alhazen's Theory of Vision and its Reception in the West." *Isis* 58 (1967): 321–341.

——. *John Pecham and the Science of Optics*. Madison, 1970.

Litt, Thomas. *Les Corps célestes dans l'univers de Saint Thomas d'Aquin*. Louvain and Paris, 1963.

Little, A. G., ed. *Roger Bacon Essays*. Oxford, 1914.

McColley, Grant. "An Early Friend of the Copernican Theory: Gemma Frisius." *Isis* 26 (1937): 322–325.

McKie, Douglas, and N. H. de V. Heathcote. *The Discovery of Specific and Latent Heats*. London, 1935.

McMullin, Ernan, ed. *Galileo, Man of Science*. New York and London, 1967.

McTighe, Thomas P. "Galileo's 'Platonism': A Reconsideration." In *Galileo, Man of Science*. E. McMullin, ed. Pp. 365–387.

Martin, Th. H. "Sur les instruments d'optique faussement attribués aux anciens par quelques savants modernes." *Bulletino di bibliografia e di storia delle scienze mathematiche e fisiche* 4 (1871): 165–238.

Mélanges Alexandre Koyré, publiés à l'occasion de son soixante-dixième anniversaire. 2 vols. Paris, 1964.

Mélanges d'histoire offerts à Henri Pirenne par ses anciens élèves et ses amis à l'occasion de sa quarantième année d'enseignement à l'Université de Gand, 1886–1926. 2 vols. Paris, 1926.

Mercator, Gerardus. *Atlas sive cosmographicae meditationes de fabrica mundi et fabricati figura*. Duisburg, 1595(?).

——. *Correspondance mercatorienne*. M. van Durme, ed. Antwerp, 1959.

Michel, Henri. *Scientific Instruments in Art and History*. R. E. W. Maddison and F. R. Maddison, trs. London, 1967.

Mizauld, Antoine. *Planetologia, rebus astronomicis, medicis, et philosophicis erudite referta*. Lyon, 1551.

Montaureus, P. *See* Euclid. *Euclidis elementorum*.

Naudé, Gabriel. *Apologie pour les grands hommes soupçonnez de magie.* Amsterdam, 1712.

Nauert, Charles G., Jr. *Agrippa and the Crisis of Renaissance Thought.* Urbana, 1965.

Nordenskiöld, A. E. *Facsimile-Atlas to the Early History of Cartography.* J. A. Ekolof and C. R. Markham, trs. Stockholm, 1889.

———. *Periplus. An Essay on the Early History of Charts and Sailing-Directions.* F. A. Bather, tr. Stockholm, 1897.

Nuñez, Pedro. *De crepusculis.* Lisbon, 1542.

Ortroy, F. van. "Bio-bibliographie de Gemma Frisius, fondateur de l'École belge de géographie, de son fils Corneille et de ses neveux les Arsenius." Brussels. Académie Royale de Belgique. Classe des Lettres et des Sciences Morales et Politiques. *Mémoires in-8°,* 11:2 (1920), 418 pp.

———. "Les sources scientifiques de la cartographie mercatorienne." *Mélanges d'histoire offerts à Henri Pirenne.* Vol. 2. Pp. 635–652.

Paracelsus, Theophrastus Bombastus von Hohenheim. *The Hermetical and Alchemical Writings.* A. E. Waite, tr. London, 1894.

Patterson, Louise Diehl. "Recorde's Cosmography, 1556." *Isis* 42 (1951): 208–218.

Peregrinus, Petrus. *De magnete, seu rota perpetui motus, libellus.* Achilles P. Gasserus, ed. Augsburg, 1558.

Proclus. *Les Commentaires sur le premier livre des Éléments d'Euclide.* Paul ver Eecke, tr. and ed. Bruges, 1948.

Ptolemy, Claudius. *Almagest.* R. Catesby Taliaferro, tr. In *Great Books of the Western World,* vol. 16, pp. 5–465. Chicago, 1952.

———. *Cl. Ptolemaei pelusiensis mathematici Operis quadripartiti in latinum sermonem traductio. Adiectis libris posterioribus, Antonio Gogava graviense interprete.* Louvain, 1548.

———. *Tetrabiblos.* F. E. Robbins, tr. Cambridge, Mass., and London, 1940. (Loeb Classical Library, no. 350.)

Ramus, Petrus. *Praefationes, epistolae, orationes.* Marburg, 1599.

Ramusio, Giovanni Batista. *Navigationi e viaggi.* 2nd ed. 3 vols. Venice, 1554–1559.

Recorde, Robert. *Pathway to Knowledge.* London, 1551.

Reuchlin, Johann. *De arte cabalistica.* Hagenau, 1517.

Rice, Eugene F., Jr. *The Prefatory Epistles of Jacques Lefèvre d'Étaples and Related Texts.* New York and London, 1972.

Righini Bonelli, M. L., and William R. Shea, eds. *Reason, Experiment and Mysticism in the Scientific Revolution.* New York, 1975.

Ronchi, Vasco. *Optics: The Science of Vision.* Edward Rosen, tr. New York, 1957.

Rose, Paul Lawrence. "Commandino, John Dee, and the *De super-ficierum divisionibus* of Machometus Bagdedinus." *Isis* 63 (1972): 88–93.

Rosen, Edward. "John Dee and Commandino." *Scripta Mathematica* 28 (1970): 321–326.

Ross, Richard P. *Studies on Oronce Finé (1494–1555)*. Unpublished Ph.D. thesis. Columbia University, 1971.

Rossi, Paolo. *Francis Bacon: From Magic to Science*. Chicago, 1968.

———. "Hermeticism, Rationality and the Scientific Revolution." *Reason, Experiment and Mysticism in the Scientific Revolution*. M. L. Righini Bonelli and W. R. Shea, eds. Pp. 247–273.

Rulandus, Martinus. *Lexicon of Alchemy*. A. E. Waite, tr. London, 1964.

Schott, Gaspar. *Magia universalis naturae et artis*. 4 vols. Bamberg, 1677.

Scriba, Christoph J. "The Autobiography of John Wallis, F. R. S." London. Royal Society. *Notes and Records* 25 (1970): 17–46.

Sepher Yetzirah. *See* Joseph, Rabbi Akiba ben.

Settle, Thomas B. "Galileo's Use of Experiment as a Tool of Investigation." In *Galileo, Man of Science*. E. McMullin, ed. Pp. 315–337.

Shumaker, Wayne. "Accounts of Marvelous Machines in the Renaissance." *Thought* 51 (1976): 255–270.

———. *The Occult Sciences in the Renaissance*. Berkeley, 1972.

Simpkins, Diana M. "Early Editions of Euclid in England." *Annals of Science* 22 (1966): 225–249.

Skelton, R. A. "Mercator and English Geography in the 16th Century." *Duisburger Forschungen* 6 (1962): 158–170.

Smet, Antoine de. "Gaspar a Mirica." *Der Globusfreund* 13 (1964): 32–37.

———. "Les géographes de la renaissance et la cosmographie." Brussels. Université libre. *L'Univers*. Pp. 13–29.

———. "Gerard Mercators wetenschappelijke, technische en kartografische activiteit." Brussels. Société Belge d'Études Géographiques. *Bulletin* 32 (1963): 31–49.

———. "Mercator à Louvain (1530–1552)." *Duisburger Forschungen* 6 (1962): 28–90.

Smith, Charlotte Fell. *John Dee (1527–1608)*. London, 1909.

Smith, David Eugene. *History of Mathematics*. 2 vols. Reprint, New York, 1958.

Smith, G. C. Moore. *Gabriel Harvey's Marginalia*. Stratford-upon-Avon, 1913.

Smith, Thomas. "Vita Joannis Dee mathematici Angli." In Smith, *Vitae quorundam eruditissimorum et illustrium virorum*. Pp. 1–102. London, 1707.

Sorel, Charles. *See* De L'Isle, Le Sieur.

Stadius, Joannes. *Ephemerides novae.* Cologne, 1556.

Steele, Robert R. "Meeting of Cardano and Dr. Dee." *Notes and Queries,* ser. 8, 1 (1892): 126.

Stephens, J. E., ed. *Aubrey on Education.* London, 1972.

Strong, Edward W. *Procedures and Metaphysics.* Berkeley, 1936.

Strype, John. *The Life of the Learned Sir John Cheke.* 2nd ed., Oxford, 1821.

Taylor, E. G. R. "John Dee and the Map of Northeast Asia." *Imago Mundi* 12 (1955): 103–106.

———. "John Dee and the Nautical Triangle." *Journal of the Institute of Navigation* 8 (1955): 318–325.

———. "A Letter Dated 1577 from Mercator to John Dee." *Imago Mundi* 13 (1956): 56–68.

———. *Tudor Geography.* London, 1930.

Teixera da Mota, Avelino. *Mar, além mar. Estudos e ensaios de história e geografia.* Vol. I. 1944–1947. Lisbon, 1972.

Thomas, Keith. *Religion and the Decline of Magic.* New York, 1971.

Thomas, W. Gwyn. "An Episode in the Later Life of John Dee." *Welsh Historical Review* 5 (1971): 250–256.

Thomas-Stanford, Charles. *Early Editions of Euclid's Elements.* London, 1926. (London. Bibliographical Society. *Illustrated Monographs* 20.)

Thomas Aquinas, Saint. *Commentary on Aristotle's Physics.* Richard J. Blackwell, Richard J. Spath, and W. Edmund Thirkel, trs. London, 1963.

———. *Summa contra gentiles.* 4 vols. New York, 1955–1956.

Thorndike, Lynn. *History of Magic and Experimental Science.* 8 vols. New York, 1923–1958.

Tilley, Arthur. "Greek Studies in England in the Early 16th Century." *English Historical Review* 53 (1938): 221–239, 438–456.

Tillyard, E. M. W. *The Elizabethan World Picture.* London, 1950.

Trattner, W. I. "God and Expansion in Elizabethan England: John Dee, 1527–1583." *Journal of the History of Ideas* 25 (1964): 17–34.

Trithemius, Joannes. *Steganographia.* Frankfurt, 1606.

Turetsky, M. "Permutations in the 16th Century Cabala." *Mathematics Teacher* 16 (1923): 29–34.

Turner, Robert. *Ars Notoria: The Notory Art of Salomon.* London, 1657.

Tyard, Pontus de. *Mantice, ou Discours de la verité de divination par astrologie.* 2nd ed., Paris, 1573.

Urbanitzky, Alfred von. *Elektricität und Magnetismus im Alterthume.* Vienna, 1887. (Reprinted, Wiesbaden, 1967.)

Urso Salernitanus. *Die Medizinisch-naturphilosophische Aphorismen und Kommentar.* R. Creutz, ed. Berlin, 1936. (*Quellen und Studien zur Geschichte der Naturwissenschaften und der Medizin* 5:1.)

Verheiden, W. J. *Vita Guillelmi Verheiden*. Leyden, 1596.

Vogl, Sebastian. *Die physik Roger Bacos*. (*13 Jahrh.*). Erlangen, 1906.

———. "Roger Bacons Lehre von der sinnlichen Spezies und vom Sehvorgange." *Roger Bacon Essays*. A. G. Little, ed. Pp. 205–227.

de Waard, Cornelius, Jr. *De uitvinding der verrekijkers: Eene bijdrage tot de beschavingsgeschiedenis*. The Hague, 1906.

Walker, D. P. *Spiritual and Demonic Magic from Ficino to Campanella*. London, 1958. (Reprinted, University of Notre Dame Press, 1975.)

Waters, D. W. *The Art of Navigation in Elizabethan and Early Stuart Times*. New Haven, 1958.

Webb, Henry J. *Elizabethan Military Science*. Madison, 1965.

Webster, John. *Academiarum Examen, or The Examination of Academies*. London, 1654. (Reprinted in Allen G. Debus. *Science and Education in the 17th Century. The Webster-Ward Debate*, pp. 67–192. London and New York, 1970.)

Wedel, Theodore Otto. *The Medieval Attitude Toward Astrology*. New Haven, 1920. (Reprinted, Archon Books, 1968.)

Westman, Robert S., ed. *The Copernican Achievement*. Berkeley and Los Angeles, 1975. (UCLA Center for Medieval and Renaissance Studies. *Contributions 7*.)

———. "Three Responses to Copernican Theory: Johannes Praetorius, Tycho Brahe, and Michael Maestlin." In Westman, R. S., ed. *The Copernican Achievement*. Pp. 285–345.

Whittaker, E. T. *A History of Theories of Aether and Electricity*. Vol. I: *The Classical Theories*. 2nd ed., London, 1951.

Wiedemann, E. "Roger Bacon und seine Verdienste um die Optik." *Roger Bacon Essays*. A. G. Little, ed. Pp. 185–203.

———. "Zur Geschichte der Brennspiegel." *Annalen der Physik* 39 (1890): 110–130.

Yates, Frances A. *Giordano Bruno and the Hermetic Tradition*. Chicago and London, 1964.

———. "The Hermetic Tradition in Renaissance Science." *Art, Science, and History in the Renaissance*. Charles S. Singleton, ed. Baltimore, 1968. Pp. 255–274.

———. *The Rosicrucian Enlightenment*. London and Boston, 1972.

———. *Theatre of the World*. Chicago, 1959.

Zubov, Vasilii Pavlovich. "La Formule calorimétrique et ses origines." *Mélanges Alexandre Koyré*. Vol. 1. Pp. 654–661.

———. "Kalorimetricheskaia formula Rikhmana i ee predistoriia." Akademiia nauk S.S.S.R. Institut istorii estestvoznaniia i tekhniki. *Trudy* 5 (1955): 69–93.

❧ Index

Abat, Bonaventure, on magnifying mirrors, 71
Abulfeda (Abu'l-Fidā' Ismā'il), 14
Accademia del Cimento, 39
Accademia del Disegno, 47
accident, defined, 207
action at a distance, 61
Acts 4: 25, cited, 234
agent, feeble, as sometimes powerful, LXXVII (163)
Ages of the Father, Son, and Holy Spirit, 238
Agrippa von Nettesheim, Heinrich Cornelius: on abstractions of astronomers, 98; astrology, 41; conjuring, 40; numerology, 212–213; planetary sounds, 209; vacillations, 43; mentioned, 52
Alberti, Leone Battista, 47
Albertus Magnus, Saint, 33
Alchabitius, 53, 97
alchemy: alluded to, LXXVII (163–165); importance of sympathies to, 208; purgations in, 230–231
Alfraganus on stellar sizes and planetary distances, 87
algebra: Dee's approach to, 20–22, 25–26; mentioned, 53
Alhazen, 58, 62, 67, 68
Alkindus, *see* al-Kindī
Almagest, of Ptolemy, alluded to, 46, CVIII (187), 238
Anatomical Magic, *see* Magic, Anatomical

anthropographie, defined, 33
antipathies, XII (127), 209
Apollonius of Perga, 46
applied mathematics, depreciated, 4
apsides, defined, 79
archemastrie, 33, 93 (n. 89)
archetypal ideas, 215
Archimedes: fables about, 69–70; and legitimation of mathematics, 47; mentioned, 24, 26, 53
architecture, defined, 32
Archytas's dove, 33
Aristophanes, 33
Aristotle: on efficient causes, 61; mentioned, 43, 53, 55, 92 (n. 85)
arithmetic, as highest form of mathematics, 18
ars notaria or *notoria*, *see* notary art *and* notory art
aspects, defined, 90, 229
associations, astrological, 216
Assyro-Babylonians, 229
astrological influences, *see* influence, astrological
astrology: as applied mathematics, 69, 88–90, 92–94; as physics, 43, 93 (n. 89); as understood by Cardano, 51–52, by Dee, 51, by al-Kindī, 53, by Louvain group, 54–56, by Offusius, 54, by Ptolemy, 53; and magic, 42–43; in medicine, 55, 59 (n. 32); beneficent influences, 56; conjunctions, 90–91; defended, 59; determinism, 41–42;

Rossi, Paolo, on Francis Bacon, 37–39
Rulandus, Martinus, 212

St. John's College, Cambridge, 1
schematic and aschematic intervals, defined, 234
Schott, Gaspar, 37, 70
Scot, Michael, 33 (n. 105)
seals: differences in the imprintation of, XXVI (135); stars and celestial powers as, XXVI (135)
seasons: comparison of, to a day and a lifetime, 237; length of, 79–80; similarity of annual and daily, CVII (185)
seeds, generational, XXI (131)
Sefer Yezirah, 212
sextile aspect, defined, 90
Shumaker, Wayne, 208, 209, 229, 241
sidereal period: of moon, 80; of planets, 81
sidereal year, 227; solar, defined, LXVII (157); duration of, LXVII (157)
Sidney, Philip, 9 (n. 27)
significator, defined, 94; discussed, LXXIIII (161); proper, LXXIII (159)
signs of the zodiac, *see* zodiac
sizes of celestial bodies: importance of knowing, XXX (137); of stars, 87–88
solar periods, *see* period; periods
Solomon, King, 13
solstices, defined, 76–77
Sorel, Charles, *see* L'Isle, Le Sieur de
soul, human, as possessing greater virtues than body or matter, CX (187)
species, emission of, V (123–125), 209; by immaterial substances, V (123–125); species of a species as responsible for twilights, XLVIII (147); spiritual, XIIII (127–129). *See also* multiplication of species
sphere, "perfection" of, 209
squaring the circle, *see* circle
stars: as including planets, 217; as the cause of long-term events, LXXVI (163); direction and progression of, defined, 90; scale of brightness of, 88 (n. 76); sizes of, 87–88, LXXXVIII (165), 231; causes assisting influence of, LXXVIII (165); inalterable rela-

tionships of, LXXV (161–163); larger than the earth, rays from, XXXVI (139); smaller than the earth, rays from, XXXV (139); rays from, before rising or setting, XLVI (145); names of, as derived from gods, L (147); the precession of, LXXV (161–163); prestige of, 210
statike, defined, 32
Statius, Joannes, 56, 57
Stephanus Gracilis, 7, 18
Stifel, Michael, 21 (n. 68), 48
Strong, Edward W., 48
sublunary sphere, imperfection of, 210
substance, defined, 207
sun: apparent speed of, along the ecliptic, 79–80; as chief source of light and heat, XCV (179); distance from earth of, 82–84; in Capricorn and Cancer, XLVII (145), 222; eccentric orbit of, 79–80; observations of, 78–79; as principle of heat, 92–93; as surpassing other celestial bodies in light, CII (183); twenty-six positions of, 238; cause of procreation and preservation, CVI (185)
Sylvester, Pope, 33 (n. 105)
sympathy, as related to correspondence, 208
sympathy and antipathy, XII (127)
synodic period: of moon, 80; of planets, 81

talismans, 94, XXVI (135), 208, 215–216, 239
tenth house, cusp of, 235
thaumaturgike, defined, 33
Thebites, LXVII (157)
thermometers, invention of, 235
Thomas Aquinas, St.: on astrology, 41–42; on conjuring, 40; on magnetism, 62 (n. 38)
Thorndike, Lynn, 37
thoughts, as obedient to bodies, XXIII (133)
Tillyard, E. M. W., 229
trine aspect, defined, 90
Trinity, physical mysteries in, CVII (187)
Trinity College, Cambridge, 33
Trithemius, Joannes, Abbot of Sponheim, 18, 52, 213